Schizophrenia Genesis

A Series of Books in Psychology

Editors: Richard C. Atkinson
Gardner Lindzey
Richard F. Thompson

Schizophrenia Genesis
The Origins of Madness

Bruce Africa
6/10/91

Irving I. Gottesman
University of Virginia

With the Assistance of
Dorothea L. Wolfgram

W. H. Freeman and Company
New York

Library of Congress Cataloging-in-Publication Data

Gottesman, Irving I.
  Schizophrenia genesis : the origins of madness / Irving
  I. Gottesman with the assistance of Dorothea L. Wolfgram.
  p.   cm. — (A Series of books in psychology)
  Includes bibliographical references.
  Includes indexes.
  1. Schizophrenia—Etiology.   I. Wolfgram, Dorothea L.   II. Title.
III. Series
  [DNLM: 1. Schizophrenia—etiology.   WM 203 G685sb]
RC514.G673   1991
616.89′82071—dc20
DNLM/DLC   90-3840
for Library of Congress   CIP
ISBN 0-7167-2145-7
ISBN 0-7167-2147-3 pbk.

Printed in the United States of America

1 2 3 4 5 6 7 8 9 0 VB 9 9 8 7 6 5 4 3 2 1

To Carol, Judy, David, and Adam
for freeing me to think and write

Contents

Preface

Everyday misuse of the terms *schizophrenia* or *schizophrenic* to refer to the foreign policy of the United States, the stock market, or any other disconfirmation of one's expectations does an injustice to the enormity of the public health problems and profound suffering associated with this most puzzling disorder of the human mind. Current surveys in the United States indicate that 1.85 million Americans over the age of 16 have or have had an episode of schizophrenia. By the year 2000, that number will grow to 2.06 million victims. Worldwide projections are quite close to rates in the United States, whether they come from rural India or industrialized England, Japan, or Denmark, according to recent reports from the World Health Organization.

Sorting out fact from fiction and myth from reality in regard to the causes and origins of a major mental disorder such as schizophrenia is no easy task for either a mental health professional or a curious layperson, whether they are the relative or friend of a human being suffering from schizophrenia or not. A heritage of distortions, stagnant certainty, and self-serving territoriality characterizes the fields of knowledge about this dreaded disorder — aptly called "the cancer of the mind."

The goal of this book is to make the complex information about schizophrenia, which comes from diverse sources and cuts across the boundaries of traditional scientific domains, accessible and comprehensible to an audience wider than the one facing a lecturer in an advanced university course in abnormal psychology or a medical school course on the origins of mental illness. I offer this telling of the schizophrenia story, enlivened by case histories of patients and their families in the oral tradition, to help fill the information gap between the "ivory tower of academia," with its research "factories" and private language, and the idiosyncratic narratives glorifying *or* obfuscating disorders of the mind. My viewpoint is necessarily constrained by the range of my expertise in academic and clinical psychology, clinical psychiatric genetics, and social biology. Thirty years of personal, research, and clinical experience with schizophrenia do not guarantee clarity of communication to a diverse audience. But, I hope that both friends and relatives of the mentally ill and my fellow scientists and academicians, all of whom strive for a greater understanding of the causes and the consequences of mental illness, may be served by this book. I have tried to simplify without oversimplification; to explain with a minimum of tabular presentations, figures, and drawings; and to provide an entré to the scientifically detailed literature for those who wish to pursue the discussion of any particular topic in more depth.

The telling of the story is complete with the warts and seams of the processes by which scientists acquire and promulgate the facts and theories that serve as a bridge to their subsequent efforts to refute or confirm their views. The scientists are revealed as the fallible, egoistic, political, territorial, and *humane* beings that they are. Science, unlike many other human endeavors, is a self-correcting process when given half a chance and an unfettered political and economic climate. We can expect the search for the causes of schizophrenia to be accompanied by the highs of discovery and confirmation as well as the lows and embarrassment of refutation. My scientific and philosophical orientation toward schizophrenia is very close to that of those who pursue the causes of diabetes and coronary heart disease, but without slavish loyalty to a so-called medical model. Schizophrenia is a complex disorder of human functioning. The absence so far of a solution to its origins compels me to be skeptical about received wisdom from all participants, however noble and well-intended. I am, however, optimistic about finding solutions via the energies of scientists and the canons of science within a decade.

I wish to acknowledge my debt to the Center for Advanced Study of the University of Virginia for a sabbatical leave, to the Center for Advanced Study in the Behavioral Sciences, Stanford, California, for an unfettered climate during the 1987–1988 academic year, and to the John D. and Catherine T. MacArthur Foundation and the Scottish Rite Schizophrenia Research Program (Northern Masonic Jurisdiction) for financial assistance in my research and writing. I want to thank Kathleen Much of the Center for Advanced Study in the Behavioral Sciences for her efforts to make my prose intelligible and Jonathan Cobb, Diana Siemens, and Diane Maass of W. H. Freeman and Company for their detailed editorial advice. I must thank my peers during my year at the Center in Stanford for demanding clarification of matters *I* thought *were* clear, as well as my friends and colleagues who continue to share my enthusiasm for learning about schizophrenia and to further my education: Aksel Bertelsen, Manfred Bleuler, Nikki Erlenmeyer-Kimling, Anne Farmer, Leonard Heston, Daniel Hanson, Matt McGue, Peter McGuffin, Paul Meehl, and

Fuller Torrey. I am grateful to Mary Ellen Peters, Margaret Schneyer, and Carol Prescott for their help in the production of the manuscript.

Irving I. Gottesman
Charlottesville, Virginia
July 1990

Schizophrenia Genesis

1

In the Beginning
Schizophrenia in History

If madness is as old as humankind, we might be tempted to assume that schizophrenia, one of today's best known and most common forms of madness, has been present since the dawn of civilization. No conclusive proof of that assumption exists, however. We have found no ancient Sumerian or Babylonian cuneiform describing an insanity that arises in adolescence, causes hallucinations and delusions, and eventually goes away, but often recurs — the hallmark symptoms of schizophrenia. In ancient writings, we find no description of mental illness that is, by contemporary standards, unquestionably schizophrenia, though there are descriptions that would fit modern diagnoses of senile dementia and severe depression. These facts lead some distinguished authorities to suggest that schizophrenia was rare, or even nonexistent, before the nineteenth century.

It may be hard to imagine such a striking illness springing up from nowhere less than 200 years ago, but nineteenth-century writers in Europe did report an alarming, acute increase of unmanageable insanity in their societies. Consider, however, the recent and apparently acute ap-

pearance of AIDS (acquired immunodeficiency syndrome), the end stage
of infection with HIV (human immunodeficiency virus). No cases were
observed anywhere in the world until 1977–1978. When formal surveil-
lance began in the United States and Europe in 1982, there were fewer
than 1,000 known cases in the United States. By early 1990, the number
had grown to 115,000, with 435,000 cases projected by the end of 1993.
New diseases can and do arise; discovering their origins and causes
requires creative research strategies.

The general historical observations with respect to schizophrenia go
something like this: There is no ancient description of schizophrenia,
although early healers, including the Father of Medicine, Hippocrates
(460–377 B.C.), a contemporary of Socrates, carefully described other
forms of madness such as the "sacred disease" (epilepsy), mania, and
melancholia (depression). At least Hippocrates attributed madness to the
brain — as opposed to possession by the gods, the conventional classical
explanation for bizarre behavior — but thought it was due to "abnormal
moisture" of the brain. Once schizophrenia was adequately described
clinically in 1809, however, the disease seems to have become visible all
over the Western world and to have increased rapidly for a hundred
years. If this observation is valid, it demands an explanation.

One theory holds that a tendency toward this kind of mental breakdown may have always been present in humans, but emerged as an
incapacitating illness only under the stress of losing personal space in
increasingly urbanized and industrialized societies. Although no one has
ever ruled out stress as a contributing factor, it now seems clear that it
does not cause schizophrenia by itself.

Closely allied, but slightly different, is a second theory: The social
changes caused by the breakdown of traditional family and cultural
patterns — patterns that had remained much the same from ancient
times up to the industrial revolution in the eighteenth century — created
more insanity of all types, and thus more schizophrenia. As more and
more insane asylums, or madhouses, were built in western Europe during
the nineteenth century, these concentrated gatherings enabled observers
to differentiate carefully, for the first time, among various forms of
madness, including schizophrenia. Only with the coming of these asylums were physicians able to define different mental illnesses by describing the differences in symptoms, in course, and in outcome, because only
then could they observe large enough numbers of insane persons for long
enough periods of time.

Much simpler, but also more radical, is the third theory: Schizophrenia
is a disease that actually did not exist before the seventeenth or eighteenth century. That's why no one had described it earlier and why

nineteenth-century observers saw such an increase in this type of insanity. It came and it spread, perhaps transmitted by an infectious viral agent, paralleling the AIDS story. Perhaps schizophrenia was spread like the psychoses caused by syphilitic invasion of the central nervous system; that condition — called general paralysis of the insane (GPI), general paresis, or dementia paralytica — may have resulted from a mutated virus that appeared in France soon after the Napoleonic wars and spread throughout the West.

Each of these three theories to account for the sudden appearance of large numbers of sufferers from schizophrenia, however, must accommodate the fact that, since about 1900, there has been no notable increase in the incidence of schizophrenia anywhere in the world. Industrialization has spread; family breakdowns have, if anything, increased; and no antischizophrenic viral agent has been invented.

Figure 1 shows the admission rates for all "lunatic asylums" for all conditions in England and Wales from 1860 to 1914. Clearly, something had changed, but what? Better reporting, or more lunatics, or . . . ? The period from 1840 to 1910 was known as the "asylum era" — the number of British institutions built during those years mushroomed from about 20 to 90. But the data in Figure 1 are ambiguous sources of information about changes in the frequency of schizophrenia per se, since the numbers refer to all varieties of "lunacy." Because no central reporting of

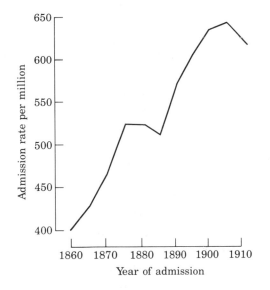

FIGURE 1.   Insane asylum admission rates per one million population in England and Wales for the period 1859 to 1914. Adapted from Hare (1983b).

insanity was required until the Lunacy Act of 1845 (which covered England and Wales), estimates of lunatics are unreliable. "Pauper lunatics," for example, who had been lumped in with other vagrants and indigents in the workhouse statistics, were transferred onto the rolls of the lunatic asylums after 1845. It is difficult today to discover exactly where the mentally retarded and the elderly with senile dementia were counted before and after the 1845 legislation.

Since schizophrenia was not reliably diagnosed and separated from other forms of psychosis until the very end of the nineteenth century, what reasoning about its apparent scarcity in earlier records seems sound? It may have been present but infrequent because demented people were less likely than normal ones to survive in a harsh, primitive society. A few slightly mad and cunning people might have been valued as seers, magicians, witch doctors, or religious leaders, but data about both their diagnoses and their fertility would require sheer speculation. It's unlikely, however, that even a Darwinian weeding out of schizophrenics would have wiped out the disorder. As we will see in subsequent chapters, persons who do not exhibit schizophrenia themselves can carry an inherited predisposition to it. And two parents, with this predisposition, even though they show no symptoms at all, can pass along to their children a much greater tendency to schizophrenia than would either parent alone. Furthermore, schizophrenics who have children almost always have them before they become ill. No Darwinian natural selection can eliminate a disease from our species under such conditions. Recent research suggests that, even today, some process of natural selection against successful reproduction is at work among schizophrenics. They are not replacing themselves in the population (perhaps they never did), but the incidence of the disorder is not decreasing conspicuously.

Today we find human beings afflicted with schizophrenia in all societies and in all socioeconomic strata within these societies. Overall, about 1 in 100 persons will develop this condition by age 55 or so. Differences in frequency over time and space (cultures, subcultures) have sometimes been observed, but they are difficult to interpret and will be examined in more depth in Chapter 4. Many of the reported variations must be artifacts of inaccurate reporting (untrained observers, small samples, idiosyncratic conventions about diagnoses), and perhaps simpler explanations relating to changes in life expectancy and to social mobility will suffice. Because evidence about the appearance and spread of schizophrenia is so important in guiding strategies for detecting the causes, let us look briefly at attempts to document the existence of schizophrenia before a scientific approach to psychiatry and mental illness was developed.

Early Accounts of Schizophrenia

The ancient texts combed most thoroughly for descriptions of something akin to schizophrenia are, predictably, the Old and New Testaments. The prophet Ezekiel, for instance, has been "diagnosed" as schizophrenic by modern writers because of his ecstatic visions. The description of King Saul in the First Book of Samuel suggests severe mental illness, but it strikes a modern reader more as a description of psychotic depression with paranoid delusions. Early in the twentieth century, a rash of psychobiographies claimed to "document" Jesus's schizophrenia. Each diagnosis rests on Jesus's delusion of grandeur—believing he was God's son—and on his auditory hallucinations—talking with God and other figures. Dr. Albert Schweitzer issued a rebuttal in his 1913 doctoral thesis, *The Psychiatric Study of Jesus*. He found the numerous publications (going back to 1835) to be both historically and clinically unsound and dismissed them as ethnic and religious slurs.

Later documents also provide little solid evidence of schizophrenia. Descriptions of 57 "chronicles of miracles" at Saint Bartholomew's Hospital in London (the "mother hospital" of the English-speaking world), translated from Latin manuscripts in the twelfth century, reveal only four possible cases of schizophrenia, having both auditory and visual hallucinations, delusions, and frequent recovery—classic characteristics of the disorder. No unambiguously schizophrenic character appears in Shakespearean drama, despite the Bard's skill as a word-painter of other kinds of behavioral deviances. Careful notes and symptom checklists compiled by the seventeenth-century English physician-astrologer Richard Napier (1559–1634) on more than 2,000 mental patients do not reveal any clear cases of schizophrenia. The term *neurosis* was not introduced until 1783, when it referred generically to a disordered nervous system in the absence of fever; *psychosis* did not appear as a term until the middle of the nineteenth century. Still, we cannot find unmistakable reports of schizophrenia under these or other names in early medical accounts.

The first clinically adequate descriptions of schizophrenia appeared independently in England and France in the same year, 1809. In *Observations on Madness and Melancholy*, John Haslam (1764–1844), superintendent of the Bethlem Hospital in London, described what we recognize today as unmistakable schizophrenia:

There is a form of insanity which occurs in young persons; and as far as these cases have been the subject of my observation, they have been more frequently noticed in females. Those whom I have seen, have been distinguished by prompt capacity and lively disposition; and, in general have become the favorites of parents and tutors, by their facility in acquiring knowledge, and by a prematurity of attainment. This disorder commences, about or shortly after, the period of menstruation, and in many instances has been unconnected with hereditary taint, as far as could be ascertained by minute inquiry. The attack is almost imperceptible; some months usually elapse before it becomes the subject of particular notice; and fond relatives are frequently deceived by the hope that it is only an abatement of excessive vivacity, conducting to a prudent reserve, and steadiness of character. A degree of apparent thoughtfulness and inactivity precede, together with a diminution of the ordinary curiosity, concerning that which is passing before them; and they therefore neglect those objects and pursuits which formerly proved courses of delight and instruction. The sensibility appears to be considerably blunted: they do not bear the same affection towards their parents and relations: they become unfeeling to kindness, and careless of reproof. To their companions they show a cold civility, but take no interest whatever in their concerns. . . . Thus in the interval between puberty and manhood, I have painfully witnessed this hopeless and degrading change, which in a short time has transformed the most promising and vigorous intellect into a slavering and bloated idiot. (pp. 64–67).

Although Haslam was wrong on two counts — overstating schizophrenia's occurrence in women and minimizing its heritability — he touched upon many of the symptoms his successors have carefully documented. He also evoked some of the tragedy inevitably encountered by those who study and treat this disease.

Also in 1809, Philippe Pinel (1745–1826), a French physician, clearly characterized cases of typical schizophrenia. During the next 50 years, more than a dozen good descriptions were published, almost all giving the disease different names. These descriptions spilled over from scientific to general literature; by 1832, the major character portrayed in Balzac's short story "Louis Lambert" is clearly schizophrenic.

In 1852, Benedict Morel (1809–1873), physician-in-chief of a French institution, first used the French term *démence précoce* (in Latin, the language of medicine, *dementia praecox*) to describe what we now know as schizophrenia. Pinel had called the illness a *démence* (loss of mind) to characterize the deterioration of mental abilities he saw in his chronically ill, hospitalized patients. Morel added *précoce* (early, premature) to identify the conspicuously frequent onset of the disease in adolescents and its "galloping" course. Morel believed that the illness was inherited and concocted a model to explain the degeneration. His claim that its cause was heredity alone was far off the mark, but his clinical description of the patient he had in mind when he coined the term was accurate. He described a boy of 13 or 14 who had seemed to be very bright in childhood and had normal mental development, although he was physically under-developed. "His brilliant intellectual faculties underwent in time a very distressing arrest. A kind of torpor akin to hebetude [dullness, lethargy] replaced the earlier activity," wrote Morel. In the hospital, the child improved physically, but worsened mentally and eventually was considered a hopeless case. Haslam's and Pinel's original observations 50 years before had noted the teenage onset, so Morel's term seemed appropriate.

Between Morel in the middle of the nineteenth century and the revolution in classifying mental disorders that was initiated by Emil Kraepelin in the Heidelberg Clinic at the end of the nineteenth century, manic-depressive insanity was well characterized by French psychiatrists as a separate and different form of psychotic disorder from dementia praecox. Other German psychiatrists described *catatonia* ("frozen" posturing and mutism punctuated with complete loss of motor control), *hebephrenia* (silly, fatuous, immature emotionality), and *paranoia* (delusions of grandeur) as separate, different forms of insanity or psychosis. Another opinion, with few followers, held that all these psychoses, including manic-depressive insanity and *démence précoce*, were really one unitary psychosis (*Einheitspsychose* in German) progressing toward deterioration. The concept was revitalized by Karl Menninger in 1963, supported by sophisticated scholarship.

Emil Kraepelin (1856–1926), the famous German clinical psychiatrist, became the definitive categorizer and organizer of the abnormal language and behavior that make up the substance of contemporary psychopathology. (Karl Jaspers, a younger contemporary of Kraepelin at the same university, complemented the unfolding understanding of schizophrenia from a phenomenological-descriptive angle by specifying the pathology of mental processes.) Kraepelin's clinical analyses of schizophrenics stand, for the most part, as the descriptive terms used today. He unified the formerly distinct categories discussed above and gave this comprehensive malady the name *dementia praecox*. Kraepelin believed that an adolescent who developed hallucinations and delusions, behaved bizarrely, and remained ill in this way for an extended period of time suffered from the illness Morel had described under this rubric, but he also included patients suffering from catatonia and hebephrenia. Kraepelin defined the disease as a series of clinical pictures with a common feature: termination in "mental weakness." He also admitted that the deterioration did not always occur, that the patients did not all remain chronically ill, and that dementia sometimes began after adolescence.

After long and careful observation, Kraepelin classified his dementia praecox patients according to two criteria: their symptoms and the course of their illness. By 1896, often using terms borrowed from others, he divided his patients into three types: *hebephrenic, catatonic,* and *para-noid.* In 1913, he added dementia *simplex* (those patients with only subtle, "negative" symptoms) to his classificatory scheme. These descriptions not only characterized the patients' symptoms, but indicated the severity of the psychoses, moving from simple and paranoid, the least severe, to hebephrenic, the most. By measuring the severity of the patient's illness, Kraepelin also was predicting some correlation between how ill the patient was and the degree of recovery to be expected, a so-called prognosis.

Throughout the twentieth century, psychopathologists have taken issue with the term *dementia praecox.* It implied a fatalistic, gloomy outcome in all cases, a deterioration to *idiocy* (Pinel's word) or *stupidité* (Morel's). Yet physicians observed some severely "demented" patients who got well, partially or fully. It implied adolescent onset. But clinicians encountered adults at the age of 40, 50, or even 60 who became ill with dementia praecox for the first time.

In Switzerland in 1908, Eugen Bleuler (1857–1939) — a contemporary of Kraepelin, Jaspers, Freud, and Jung — introduced the term *schizophrenia*; literally, splitting of the mind. By renaming the disease to focus on a splitting of usually integrated psychic functions (that is, mental associations), Bleuler tried to call attention to the phenomena Kraepelin had downplayed — frequent social recovery and emotionality — thereby avoiding the focus on a relentlessly downhill course and hopeless outcome. He believed that a yet-to-be-determined toxin caused alterations in the thinking process and disharmony of affect among schizophrenics. (*Affect,* in the psychiatric sense, refers to how one responds to the world, one's adaptation and emotional response or "feeling tone." Thus, a patient with a *flat affect* fails to show emotional change under varying circumstances, and one with an *inappropriate affect* laughs, for instance, when speaking of the death of a loved one, or cries for no apparent reason.)

Bleuler's reconceptualization contributed greatly to the scientific understanding of schizophrenia as a disruption of the thought processes and feelings of *persons* who, in most other ways, are like ourselves. An unintended side effect of his term, however, was that many people came to believe that schizophrenics display multiple, or split, personalities. They do not. Schizophrenia can — for a time, for many times, or forever — change its victim from a rational, intelligent human being into an irrational, totally deranged person, but it does not create a new person.

Perhaps scientists would have done better to reject both Bleuler's new term and Morel's old one in favor of a name totally neutral in its implications. One historian suggests that we should have called dementia praecox/schizophrenia the Pinel/Haslam syndrome.

Historical Views of Insanity and Its Treatment: Some Misguided Clues to Causes

To better understand the history of ideas about the causes of schizophrenia, it is helpful to look briefly at how society has dealt with madness. Logically, if we know the cause of a disorder, we can suggest and devise *rational*, as opposed to merely hit-or-miss, treatments for prevention and amelioration. However, knowing what brings symptomatic relief to a sufferer from a mental disorder may provide no direct clue to the causes of the illness. For example, although aspirin is well known for its relief of headache, headache is not caused by a shortage of aspirin in the body. If prayer brings relief from bereavement and depression following the death of a loved one, we cannot infer that the depression was caused by a prior lack of attention to religious duties. Insanity has historically been seen both as a problem of the spirit or soul (religious) and as a problem of the body or brain (medical). Whichever causal view prevailed, at different times and in different cultures, greatly influenced the treatment of the insane and whether they were permitted to remain within their social group or were isolated and abandoned, held in awe as chosen by the gods, or punished or killed as chosen by Satan.

In Babylon and Egypt, where medicine was first systematized, the cause of insanity was believed to be possession by demons. Hebrew civilization continued to believe in such exogenous (outside) causes, including the wrath of God, and in Far Eastern and Hindu medicine, madness was believed to result from the struggle between destructive and restorative forces. Fragments of one of the Hindu *Vedas*, or "Works of Wisdom," about 1400 B.C., describe a victim of devils who "is gluttonous, is filthy, walks naked, has lost his memory, and moves about in an uneasy manner." This characterization was specifically differentiated from toxic confusions due to poison ingestion or alcohol consumption, as well as from elated phases that resemble what we now call manic-depressive illness. For this reason, this Hindu description is sometimes cited as the first recorded suggestion of what could be schizophrenia.

Views of mental illnesses as natural phenomena of endogenous (internal) cause predominated in classical times, but relapsed to the supernatural during the Middle Ages in Europe. The insights regarding mental illness became subjugated to religious dogma. Like all culture in this period — literature, art, music, social organization — good health or bad was integrated into religious life and given a supernatural explanation. Belief in possession by devils and demons was organized and systematized. Male and female demons, higher order and lower order, beset the mentally ill according to the kind and severity of the illness. By this interpretation, however, the insane were not themselves evil; they were the hapless victims of evil. They were at least treated humanely.

During the Middle Ages, care of the mentally ill became the responsibility of the community. As a result, socially incompetent persons were gathered together for the first time. In the Islamic world, a hospital for the insane was founded in A.D. 800 in Damascus. By the thirteenth century, a mental colony was established at Gheel, Belgium, where retarded and insane children were boarded with foster families. Soon, institutions making some provision for mental disorders were established in Spain (1365 in Granada) and England (Saint Mary of Bethlem, 1243). Bethlem Hospital, given to the city of London in 1547, was an enlightened institution. Care was thoughtful and kind, and when patients left the hospital to return to their families and communities, they were given arm badges for identification. So compassionate was the community to these released patients that vagrants allegedly counterfeited badges to pass themselves off as former Bethlem inmates.

The attacks on the insane by church and society that resulted in the burning of witches occurred not during the Middle Ages, but early in the Renaissance. At times during the seventeenth and eighteenth centuries, the mercies of Bethlem were turned into the chaos of Bedlam. (The alleged (Allderidge, 1985) cries of tortured and neglected patients caused such a din that Bethlem, elided to Bedlam, became synonymous with total confusion.)

It is a paradox that while the medieval period regressed to antiscientific explanations of mental illness, yet treated the ill humanely, the Renaissance's intellectual reenlightenment brought history's darkest hour in mental health treatment. From 1460 to 1680, more than 50,000 persons, including the mentally ill (not a majority) and other unfortunates, were brought to trial as witches in Europe. A Papal Bull of 1484 authorized the extermination of witches. A handbook for finding witches was produced by two Dominican monks in 1486 and was used for the next 200 years. The infamous mass trial of witches in Salem, Massachusetts, took place in 1692. At least a few of these women seem to have been mentally ill, accused by teenage hysterics.

Speculation about why the public sentiment toward the mentally ill and other unfortunates changed from kindness to persecution involves many elements, including obvious psychopathology in the witch-hunters themselves, power struggles between Protestants and Catholics, economic greed, the disempowerment of women, the expression of repressed sexuality in Christian doctrine, and the cruel scapegoating tendencies of our species when placed under stress.

From the Renaissance to the nineteenth century, patients who were institutionalized were restrained, often by chains; whipped; ill fed; unwashed; and treated with bloodletting, purgatives, and other "curative" tortures. Those who were not hospitalized wandered the countryside unattended, scorned, beaten. Johann Reil (1759–1813), a German psychiatric reformer, wrote of the late eighteenth century:

We incarcerate these miserable creatures as if they were criminals in abandoned jails, near to the lairs of owls in barren canyons beyond city gates or in damp dungeons of prisons, where never a pitying look of a humanitarian penetrates; and we let them, in chains, rot in their own excrement. Their fetters have eaten off the flesh of their bones, and their emaciated pale faces look expectantly toward the graves, which will end their misery and cover up our shamefulness. . . . The roar of excited patients and the rattle of chains is heard day and night, and takes away from newcomers the little sanity left to them. (Translated from Kraepelin, 1918, and cited in Alexander and Selesnick, 1966, p. 116)

In most of Europe, more humane treatment of the mentally ill did not come during the intellectual Enlightenment, but afterward. Pinel, as the new administrator of Bicêtre in Paris, reformed that institution in 1793. He released patients from restraints, treated them with kindness, and fed them nourishing food. Two years later, he did the same at Salpêtrière, the women's asylum. In Italy, reform began five years before Pinel's action, under the leadership of one omnipotent grand duke and his hospital director. An English Quaker merchant family, headed by William and Samuel Tuke, admirers of Pinel, opened York Retreat in 1796 to demonstrate that Pinel's humane treatment could alleviate mental illness. Hospital reform did not spread like wildfire. In the middle of the next century, social reformer Dorothea Dix (1802–1887) took up the cause in the United States, aided by changed attitudes toward suffering and social welfare, and won American hospital and prison reforms, including the establishment of Saint Elizabeths Hospital (1855) and 32 state mental hospitals — asylums, as she preferred to think of them. The first Public Hospital for Persons of Insane and Disordered Minds in North America was opened for business in colonial Williamsburg, Virginia, in 1773, with 24 rooms, thereby declaring that mental disorder was nearer to disease than to criminality, vagrancy, or godlessness.

Origins of Modern Considerations in Mental Disorders

A century elapsed between Pinel's and Haslam's first descriptions of what may have been schizophrenia and Kraepelin's important work in studying it, distinct from other severe mental disorders. Understandably controversies surrounded questions about the causes of insanity throughout the nineteenth century since the treatments described above had not led to any specific ideas about psychological and moral versus organic (involving the brain and body) origins. Psychiatry and neurology did not begin to emerge as disciplines within medicine until the 1850s. Scientific recognition of brain malfunctioning or damage as a cause of disordered behavior was slow in coming; alcohol dependence and delirium tremens with its delusions and hallucinations was described in 1813; the dementia and psychotic mentation associated with syphilis of the brain was recognized in 1822; and, paradoxically at the time, Broca's work in 1861 showed that a left frontal lobe injury led to a loss of language functions, but *not* to insanity. Ambiguity about the role of the brain versus the psychological environment in causing insanity was perpetuated by the absence of any demonstrable hole or disease in the brain of dead psychotics examined with the crude tools of the day.

By 1896, Kraepelin was observing that some 10 to 15 percent of all admissions to mental hospitals and the vast majority of all chronic patients fit the criteria he was developing for dementia praecox. He readily admitted the flaws of the term *dementia praecox*, given that nearly half of his thousand criterion cases did not become ill for the first time until they were past 25 years old, and many of his patients recovered partially or fully. While he described his patients, he also searched for a cause or causes of the "organic disease" he thought he had delineated.

He dismissed the prevailing environmental theory, that increasing urbanization and its mental strain on youth brought on the disease. To do so, Kraepelin traveled to Singapore to study firsthand a psychiatric hospital population made up of Malaysians, Javanese, and Chinese and to hear about Japanese patients; all resembling his own. From this experience, he concluded that dementia praecox in Europeans was not caused by race, climate, food, or general circumstances of life. Despite the obvious differences in urbanization and child rearing among these cultures and between these and European cultures, the illness was the same in both symptoms and outcome.

Kraepelin believed that the illness was somehow inherited. He and other central European psychiatrists alleged that from 50 to 70 percent of their schizophrenic patients had a family history of schizophrenia. Not only parents and grandparents were implicated, but brothers and sisters as well.

For these early twentieth-century researchers, no science of genetics existed to help them explain the familial patterns observed for some human characteristics. Gregor Mendel's (1822–1884) classic 1866 studies in botany to discover the mechanisms of inheritance went little noticed and were not rediscovered until the turn of this century. In his now famous experiments breeding pea plants, he inferred the transmission of simple "factors," now called genes, that controlled the red or white color of their flowers. Pure red mated with pure white always produced all red flowers in the next generation, thus establishing a sufficient "dominant" basis for red; mating second generation reds with reds, however, yielded one in four white-flowered pea plants, thus establishing a "recessive" basis for whites, requiring one white gene from each parent plant. Many fields of medicine then rushed to apply the neat principles of simple dominant and recessive heredity, with risks in siblings of 50 or 25 percent, respectively, to any disorder that seemed to be passed from one generation to the next. Experts dealing with mental illness were no exception. In 1908, A. E. Garrod's lectures before the Royal College of Physicians in London — a tour de force that pioneered human biochemical genetics — explained how inborn errors of metabolism could be accounted for by the Mendelian recessive model. The Mendelian model explained, for instance, how certain parents who appeared perfectly normal themselves could be expected to produce an albino child 25 percent of the time. Each parent must have carried a recessive gene for albinism that did not express itself because each also carried a compensatory normal gene. If two "carrier" parents produced one albino offspring, given enough such couples and large families so as to avoid sampling error fluctuations, albinism could be reliably predicted in 25 percent of their offspring.

Though the model was simple, it took nearly three decades to thoroughly unravel even a few illnesses that resulted from recessive genes. Unfortunately, schizophrenia was not one of them, nor were other major familial conditions such as alcoholism, criminality, and diabetes. Yet the alleged major role of heredity in schizophrenia was little doubted. Kraepelin, Bleuler, and even Freud, all of whom studied the disease, were in consensus.

Working with Kraepelin in Munich, Ernst Rüdin in 1916 provided the first scientifically sound genetic study of schizophrenia. Thoroughly the scientist, Rüdin examined the familial distribution of schizophrenia and concluded the obvious: No simple Mendelian dominant trait was responsible (if it were, every schizophrenic would have to have at least one affected parent). He tried the recessive model, also to no avail: Only 4.48

percent of the brothers and sisters of his schizophrenic subjects (none of whom had schizophrenic parents) could be diagnosed as schizophrenic. He did find that about 4 percent of the siblings suffered from other psychoses. However, even if he chose to call the other ill siblings schizophrenic, he could come nowhere near the 25 percent rate required for the recessive model. Rüdin tried a more complicated Mendelian model requiring the culpable genes to be at both of two independent locations (a two-locus theory). Although these theoretical predictions came closer to emprical findings ($0.25 \times 0.25 = 6.25$ percent), he was never convinced that this forerunner of a "polygenic" model fitted the data. Rüdin, who became the leader of the now-famous Munich school of psychiatric genetics, studied stepsisters and brothers and half sisters and brothers and looked at a number of environmental factors, all with negative results.

Although Rüdin was the first to invoke the idea that multiple risk factors may contribute to the development of schizophrenia, he was unable to determine how the disease was transmitted. He found that schizophrenia, other psychoses, or alcoholism in a parent increased the observed risk of schizophrenia for the sisters and brothers of schizophrenics. He and his colleagues also concluded that schizophrenia was genetically independent of typical manic-depressive disease, an illness with which it was (and is still) sometimes confused. While investigating the genetic aspects, Rüdin became thoroughly convinced that environmental factors were also critical in determining who became schizophrenic.

As a major mental illness, dementia praecox/schizophrenia attracted the attention of most psychiatrists and psychologists in the early decades of the twentieth century, including Freud and Jung. Their work on this illness is interesting historically, but because they emphasized the role of early experience, sexuality, the "return of the repressed [ideas]," and the "talking cure" in much of mental illness, their work did not contribute in any important way to further understanding of either the cause or the cure of schizophrenia. Eugen Bleuler, a teacher and colleague of Jung's, an early member of Freud's Psychoanalytic Society, and the first to introduce the term *schizophrenia*, contributed much more than a name to the study of this disease, but his important contributions in this area came more in spite of his Freudian associations than because of them.

Although Bleuler was not in basic disagreement with Kraepelin, his 1911 publication on this illness stretched Kraepelin's findings in some ways. Bleuler believed that Kraepelin's descriptions were based on secondary, rather than primary, symptoms and that the primary problem was the disordered thought process. He argued that the primary symptoms were a loosening of associations (connecting unrelated ideas, for instance), autism (complete self-centeredness), affective disturbance (inappropriate emotions and actions), and ambivalence (not being able to make up one's mind). Delusions, hallucinations, and catatonia, Bleuler believed, resulted from these more fundamental disturbances. Bleuler's 1911 text, not available in English until 1950, also showed that he was eager to take into account the psychodynamic ideas of Freud and Jung. Kraepelin was displeased, contending that Bleuler brought unnecessary interpretations to the field. He wrote in attack:

Here we meet everywhere the characteristic fundamental features of the Freudian trend of investigation, the representation of arbitrary assumptions and conjectures as assured facts, which are used without hesitation for building up of always new castles in the air ever towering higher, and the tendency to generalization beyond measure from single observations. I must frankly confess that with the best of will I am not able to follow the trains of thought of the "metapsychiatry," which like a complex sucks up the sober method of clinical observation. As I am accustomed to walk on the sure foundations of direct experience, my Philistine conscience of natural science stumbles at every step on objections, considerations, and doubts, over which the lightly soaring power of imagination of Freud's disciples carries them without difficulty. (1919/1971, p. 250)

It must be said that much of Freud's doctrine has been overly generalized by his disciples and followers beyond the conditions for which it may be of use. This has confused the study and understanding of schizophrenia by delaying biological and genetic research, and we have lived with that confusion for much of the twentieth century. Mental health professionals believe that environmental, interpersonal, and intrapsychic stressors are contributing factors; we do not believe, however, that a bad mother or father, bad mothering, or any other environmental factor alone can cause someone to become schizophrenic. We do not believe that talking to someone, in itself, can cause or cure schizophrenia.

Many scientists from various fields have studied this most puzzling illness closely for nearly 200 years. It is easy to forget that the marvelous accomplishments of our fellow human beings in art, music, literature, engineering, and architecture over this period were not accompanied by progress in understanding how the human body and brain function and malfunction. Dozens of therapies for severely disordered minds have been tried, from spinning chairs to bloodletting to psychoanalysis, to no lasting avail. It was not until the early 1950s that specifically antipsychotic drugs successfully came into use, not as a cure, but as an aid in alleviating some of schizophrenia's most troubling symptoms — the delusions, hallucinations, and disordered thinking. Neuroscience and social science have explored blind alleys, as well as many avenues that have yielded new understanding of how the brain and nervous system function and how drugs affect them. Still, we have no clear consensus among experts on the cause or causes of schizophrenia, and it does not yet have a cure. Two centuries of study have brought insights about how genetic *and* environmental factors combine to produce schizophrenia. Insight gives hope.

The chapters to follow will embrace strategies such as Sherlock Holmes might have used to trace some of the false and misleading clues, as well as the solid and overlooked evidence, toward understanding the mystery of schizophrenia and its causes. The long trail will be enlivened with narratives by patients and their families about what it means and how it feels to suffer from and with schizophrenia. Valuable leads often will be provided by the use of information gathered through *epidemiology*; that is, tracking the patterns of appearance and disappearance of schizophrenia across different family relationships, across social classes and national cultures, across time, and across differential exposure to psychosocial stressors.

If there were ever a scientific/philosophical conflict that demanded resolution, it would be the one waged over the past 200 years or so between reductionism and synthesis. The eminent evolutionary biologist, Theodosius Dobzhansky, described two approaches to the study of the functions, structures, and interrelations of all living creatures: "The Cartesian or reductionist and the Darwinian or compositionist." (1968, p. 1). This must not be taken to mean that the biological approach can be split into such categorical camps. The two aspects are each necessary and complementary. No conflict need exist, since they are simply two sides of the same coin: the search for causality. The search, as will be seen in the chapters to follow, requires a plausible and comprehensive model of how schizophrenia might be caused, treated, and prevented. Such a model will draw on the best from a number of disciplines that allow for both twisted minds and twisted molecules.

2

How Do We Know When It's Schizophrenia?

Descriptions, Definitions, and Diagnoses

To discover the causes of schizophrenia, it is essential that only those who actually suffer from this specific disease be classified, identified, or diagnosed as schizophrenics and then become the subjects of our attention. Disturbed persons who have some other condition must be excluded from further consideration to minimize the "noise" interfering without efforts to detect sometimes weak signals.

Schizophrenia is still diagnosed today as it was at the turn of the century: by its psychopathology; that is, by its abnormal patterns of thought and perception as inferred from language and behavior. We wish we had a definitive neuropathology (either chemical or anatomical) to pinpoint a valid diagnosis of schizophrenia, but that goal still eludes us. Three ingredients go into an accurate diagnosis: (1) a careful interview with the patient to determine symptoms currently present and absent, (2) a mental and physical health history of the patient, and (3) a mental health history of the patient's relatives. The first two items are essential for all diagnostic purposes; whether the third is used depends upon who is diagnosing and why. If a clinician is making the diagnosis to decide on a course of action for an individual patient, the family history is essential. If the diagnosis is made for genetic research purposes, knowledge of family history is the last thing the researcher will want; each diagnosis within a schizophrenic's family should be made "blindly" (completely without knowledge of relatedness to other family members) lest that knowledge contaminate impartial decisions about diagnosis. The research diagnostician must decide independently who, if any, among the relatives of a schizophrenic is schizophrenic. Paradoxically, what is essential for the practitioner is poor practice for the researcher.

Clinicians are aided by knowledge of the family history, even though 89 percent of all schizophrenics do not have a schizophrenic parent and 81 percent do not have a schizophrenic parent, sister, or brother. Nevertheless, years of research confirm that schizophrenia, or a tendency toward developing it, is strongly influenced by familial factors, both genetic and environmental. Therefore, finding possible schizophrenics in a person's family tree (pedigree) should lead the diagnostician to suspect and to pursue the probability that schizophrenia may well be causing the florid or the more subtle symptoms that the person reports.

From its earliest descriptions, schizophrenia was said to run in families. That tendency could implicate either genes or environment, or a combination of both. Within the past 75 years, reliable studies of schizophrenics and their relatives have established that this familial tendency is due largely to some aspect of genetic inheritance. We believe that genes in and of themselves do not result in schizophrenia, but they do establish a predisposition or substrate — an inherited vulnerability or likelihood — that, when combined with the consequences of environmental stress, tip the scale into illness. No environmental factor, by itself, has ever been proved to cause schizophrenia. In more than 100 years of study, no one has found a single case of schizophrenia caused solely by the conditions of the patient's upbringing. The alleged causal environmental conditions —such abnormal rearing factors as early death of parents, odd marriages by odd parents, and weird parent-child relations; and such adverse life events as war, poverty, disastrous interpersonal relations, and brain insults — occur very widely, but most people experiencing any of these conditions do not become schizophrenic, although they may feel distressed or experience "nervous breakdowns." During the same 100-year period, although the pattern of inheritance was not yet clear, the necessary role of genetics was established with circumstantial and indirect evidence, at least on scientific grounds. Uncertainty on the diagnostic front, however, perpetuates uncertainty about definitive causes.

There is not yet a blood test; urine or cerebrospinal fluid analysis; or CT (computerized tomography), rCBF (regional cerebral blood flow), PET (positron emission tomography), or MRI (magnetic resonance imaging) brain-imaging scan that can establish an unchallenged diagnosis of schizophrenia. New enhancements are imminent that will enable information from the electrical and magnetic fields of the brain to be used to specify the possible structural or functional abnormalities in schizophrenics. Eventually, there will be some highly specific physical method for detecting schizophrenia, as well as the predisposition to it. We know that something goes wrong chemically and/or physically in the brain of a schizophrenic, but we do not yet know what. The antipsychotic drugs used to treat schizophrenia since 1953 give clues about excesses or deficiencies of various neurotransmitter substances or receptor sites in the brain, but these are only clues. Study continues on all fronts, but no approach has yielded *the* answer.

Reducing Uncertainty in Diagnosing Schizophrenia

Clinicians experienced in the field of mental health usually agree on the presence or absence of schizophrenia in an individual patient. Yet diagnosis remains controversial because the symptoms, taken one at a time, are not unique to schizophrenia. They also can be caused by other mental or physical illnesses, by traumatic stress, by prescription medications, by street drugs, and by brain injury. An accurate diagnosis of schizophrenia thus requires a multidimensional evaluation of behaviors, some of which are shared with "normals" or with other varieties of psychiatric disorders, and some of which are relatively specific to schizophrenia. These complications alone would be enough to cause confusion. Added to them, however, are the troublesome variations in how this disorder affects individual patients.

Schizophrenic symptoms can be slightly, moderately, severely, or absolutely disabling. The first onset can appear from puberty and adolescence to well into a person's 50s and 60s. The first episode, when untreated, can last a few weeks or many years. Recovery runs the gamut from what is euphemistically called "complete social recovery" — recovery to a level tolerable in society, but probably not to the highest level of predisease functioning — to chronic, but out-of-hospital incapacitation. And the life history of a schizophrenic can include only one or two crazy episodes (mild, moderate, or severe) or a dozen or more episodes, progressively more severe and occurring at ever shorter intervals.

The variations in symptom combinations, severity, course of illness, and outcome have led some to speak of *the schizophrenias*. The argument is as old as the diagnosis. Kraepelin, unlike his contemporaries, thought he was seeing one disease "theme" with variations. His grouping of hebephrenia, catatonia, and dementia paranoias together under one entity was only later recognized as a stroke of genius. Kraepelin read through the "noise" of the varying clinical pictures and recognized one basic theme. The subtitle of Eugen Bleuler's 1911 monograph — *Dementia Praecox or The Group of Schizophrenias* — has been misinterpreted to suggest a wish to undo Kraepelin's lumping. Bleuler in fact wanted to widen Kraepelin's view to allow for a larger variety of types of illness and outcomes. His use of the plural was intended, according to his equally famous son Manfred, as a challenge to a too-ready acceptance of the idea that a single, unitary cause was adequate to explain the diverse clinical pictures observed for cases called dementia praecox. In general, the narrower, Kraepelinian criteria influenced European clinical practice and research; Bleuler's wider criteria were more often used by American diagnosticians. American psychiatry has been influenced, as was Bleuler himself, by a strong Freudian, psychodynamic approach. The narrower the criteria — say, with an insistence on the presence of certain highly discriminating symptoms, such as the belief that someone else's thoughts are being inserted into one's mind or hearing (nonexistent) voices discuss one's behavior — the fewer cases with generally deviant behavior will be diagnosed as schizophrenia. Highly discriminating symptoms are those that, when present, lead to widely agreed upon diagnoses of schizophrenia. The broader the criteria — say, poor motivation for social participation, a deterioration in personality functioning, and some disordered thought processes — the more cases with generally deviant behavior will be diagnosed as schizophrenia, including "false positive" diagnoses, because such symptoms are *not highly discriminating* for schizophrenia per se.

We believe that schizophrenia is, at its core, *essentially* one entity, one clearly definable mental disorder that must be studied in unity along the whole continuum of its manifestations. Opinions to the contrary are easily found, however. With some exceptions noted later, we are "lumpers" rather than "splitters"; resisting, for example, the pressure to construe late-onset paranoid schizophrenia as qualitatively different from earlier onset hebephrenic schizophrenia. The generic name of the disease is schizophrenia — on that mental health workers agree, just as it is agreed that the generic name for our popular frozen dessert is ice cream. Arguments persist over whether "real" ice cream is Sealtest or Häagen Dazs, or whether the most popular ice cream is chocolate, vanilla, or strawberry.

Although the Western Europeans consider the contemporary approach now taken as a standard in North America to be merely routine, the 1987 *Diagnostic and Statistical Manual of Mental Disorders — Revised* (now

DSM-III-R) has finally narrowed American diagnosis and provided a guide to train diagnosticians to a standard. The criteria outlined in the DSM-III-R and represented in structured interviews — such as the *Diagnostic Interview Schedule* and the *Present State Examination* — are basically codifications of what many "neo-Kraepelinian" clinicians have been doing for years.

The DSM-III-R lays out symptoms somewhat like specials on a Chinese menu; that is, take three symptoms from column A, one from column B, and none from column C. If the approach is mechanical, it is at least reliable, structured, and criterion based. We are better off having it than not. But now that there is a guide, we can't throw out all data and wisdom collected about schizophrenia before 1987 on the grounds that the operational criteria were not used! Valuable research began with schizophrenia family studies in 1916, and much of what Kraepelin, Bleuler, and Jaspers had to say rings true today.

According to DSM-III-R, the diagnosis of schizophrenia will be *guided* by these criteria:

A. Presence of characteristic psychotic symptoms in the active phase: any of (1), (2), or (3) for at least one week:
   (1) Two of the following:
       (a) Delusions
       (b) Hallucinations throughout the day for several days or several times a week for several weeks, each hallucinatory experience lasting more than a few brief moments.
       (c) Incoherence or marked loosening of associations
       (d) Catatonic behavior
       (e) Flat or grossly inappropriate affect (emotional tone)
   (2) Bizarre delusions (i.e., involving a phenomenon that in the person's culture would be regarded as totally implausible — e.g., thought broadcasting, being controlled by a dead person)
   (3) Prominent hallucinations (as defined in (1) (b) above) of a voice with content having no apparent relation to concomitant depression or elation, or a voice keeping up a running commentary on the person's behavior or thoughts, or two or more voices conversing with each other.
B. During the course of the disturbance, functioning in such areas as work, social relations, and self-care is markedly below the highest level achieved before onset of the disturbance.

C. Schizoaffective disorder (a combination of schizophrenic and mood symptoms) and mood disorder (depression or mania) with psychotic features have been ruled out.
D. Continuous signs of the disturbance are seen for at least six months.
E. It cannot be established that an organic factor (e.g., brain tumor or trauma, drug intoxication, etc.) initiated and maintained the disturbance.
F. If there is a history of autistic disorder (a childhood onset psychosis), the additional diagnosis of schizophrenia is made only if prominent delusions or hallucinations are also present.

For a diagnosis of schizophrenia, criteria A through F must all be met. A former arbitrary criterion, that age of onset must be 45 years or younger, has been dropped from DSM-III-R for good reason: About 10 percent of schizophrenics have their first hospital admission after age 45.

Both the DSM-III-R and the World Health Organization (WHO) International Classification of Diseases (ICD, the standard outside the United States) are helpful in drawing a diagnostic picture. According to the DSM-III-R, schizophrenia represents a group of disorders characterized by the presence of a thought disorder. The DSM directs attention to the patients' misinterpretations of reality; delusions and hallucinations; inappropriate emotional and social response; and withdrawn, regressive, or bizarre behavior. The ICD dwells more on a fundamental disturbance of personality and a frequent sense of being controlled by outside forces, as well as on the DSM's delusions, bizarre perceptions, and inappropriate emotional and social behavior. The ICD has always omitted the arbitrary DSM-III age cutoff of 45. Both classification systems assume that organic brain disease of various origins has first been excluded as a diagnostic possibility.

Despite sharing many of the same strange symptoms, there is a striking difference between the diagnostic picture of schizophrenia and that of organic psychosis: the schizophrenic patient is not disoriented, delirious, or confused. In most areas of functioning, such as shopping for food, balancing a checkbook, or nursing a baby, schizophrenics have a firm grasp on present reality; they know what day and year it is, can read the newspaper, recognize friends and relatives, and converse intelligently sometimes. Schizophrenics — unlike persons suffering from other types of dementia (those associated with clear-cut brain diseases) — maintain a clear consciousness and intellectual capacity. In fact, schizophrenic patients often believe that their behavior is rational and that others are causing a fuss over nothing. Schizophrenic mental activity and behavior, in contrast to that associated with other psychoses, can be succinctly characterized as "ununderstandable" — an observation of the brilliant psychiatrist/philosopher Karl Jaspers (1913).

Recently, Susan Sheehan, a writer on the staff of the *New Yorker*, meticulously documented several years in the life of a young New York woman suffering from schizophrenia. In *Is There No Place on Earth for Me?* Sheehan captures the pathos and the problems of the "heroine" — given the pseudonym Sylvia Frumpkin — her family, and those who wanted to help her. As Sylvia entered, left, reentered, or turned up at new clinics and hospitals, she was usually (though not always) diagnosed correctly as suffering from schizophrenia. Too often, however, if a diagnostician paid no attention to her mental health history, a misdiagnosis resulted. And this in turn sometimes resulted in incorrect and harmful treatment. When Sylvia's illness is categorized, it is called "chronic, undifferentiated schizophrenia" (a description used by DSM-III-R for severe psychotic states that recur frequently).

Sheehan's book contributes a great deal to the wider understanding of schizophrenia, schizophrenics, and their families, but Sylvia Frumpkin represents probably less than half of all persons who enter mental hospitals and are diagnosed as schizophrenic. (For many other schizophrenics whose illness is less chronic, the lifelong outlook is much more hopeful.) On both sides of the Atlantic and in most of the world, Sylvia would be classified as a "nuclear schizophrenic," one who has the first-rank symptoms (FRS) described by Kurt Schneider, a German psychiatrist influenced by Jaspers and himself very influential since the 1930s. Table 1

TABLE 1

Kurt Schneider's first-rank symptoms of schizophrenia

1. Voices speak one's thoughts aloud.
2. Two or more voices (hallucinated) discuss one in the third person.
3. Voices describe one's actions as they happen.
4. Bodily sensations are imposed by an external force.
5. Thoughts stop and one feels they are extracted by an external force.
6. Thoughts, not "really" one's own, are inserted among one's own.
7. Thoughts are broadcast into the outside world and heard by all.
8. Alien feelings are imposed by an external force.
9. Alien impulses are imposed by an external force.
10. "Volitional" actions are imposed by an external force.
11. Perceptions are delusional and ununderstandable.

lists the FRS; they are neither necessary nor sufficient for making a diagnosis of schizophrenia, but the presence of one or more of them raises an alarm signal that should not be ignored.

Because the public is often misled by popular accounts in films, biographies, and novels, it is important to recognize false examples of schizophrenia. For instance, people with multiple personalities, such as the heroine in *The Three Faces of Eve*, are not schizophrenics; they suffer from a rare form of hysterical neurosis. Mark Vonnegut (son of Kurt Vonnegut) in *The Eden Express* describes himself as having some symptoms that occur in schizophrenia — such as alterations of the senses and somatic, grandiose, religious, and persecutory delusions and hallucinations — but his clear descriptions of manic and depressive episodes come closer to meeting the criteria for bipolar affective disorder (formerly termed manic-depressive psychosis). The effects of repeated mescaline and marijuana intoxication during his hippie period virtually guaranteed hallucinations and delusions in young Vonnegut's peculiarly sensitive personality. The resulting schizophrenia-like clinical picture is bound to confuse the diagnostic picture in his case, as does his response to vitamin therapy and his "fondness" for what afflicted him. Deborah Blau, the fictional adolescent of *I Never Promised You a Rose Garden*, has some symptoms that overlap with those seen in schizophrenia (almost delusional concerns about her body and health, grandiose delusions, paranoid traits, suicide attempts), but comes closest to meeting the criteria for somatization disorder (also known as Briquet's syndrome), with her dramatic spells of blindness, deafness, paralysis, and hallucinations. Briquet's syndrome is classified with conditions formerly termed neuroses, but is probably nearer to a severe form of personality disorder.

Genuine Schizophrenia: A Case History

Compared to these examples of "false schizophrenias," what exemplifies genuine schizophrenia? The most common variety of the disorder is called *paranoid schizophrenia*. John Romano, an American clinical psychiatrist of vast experience, excerpts the following case history of a contemporary paranoid schizophrenic to illustrate the symptoms and the distress experienced by such patients and their loved ones. (We have substituted generic names for the antischizophrenic medications used in the treatment of symptoms.)

The patient — a 33 year old, white, unemployed husband and father of six living children — dates the onset of his distress to early 197[ ]. His maternal grandmother had broken her leg, and his mother had died of cancer. After the birth of a sixth child, his wife had her gallbladder as well as her remaining ovarian tube removed. After he had signed the form giving consent for his wife's sterilization procedure, he became quite upset at the prospect of having no more children as he felt this had destroyed him. He thought that because he had signed the sterilization consent form, he too had been sterilized. Since the operation, he felt his wife had less vaginal lubrication, which made it difficult for them to have intercourse. He felt the doctors had put a tube in his wife's vagina and the tube, in turn, had been introduced into his penis, which had made him impotent. Eventually, he began to smell odors, the nature of which was impossible for him to describe. But the smells would be persuasive and would in fact drive him toward or away from that which he was to do, almost as if there was an invisible wall of smell he could not penetrate. At times the messages from the smell would be similar to the impressions he would get from his boss to hurry up at work. Some of the smells were unpleasant; some were vaginal smells. Smells had powers over him.

In some way the town banker was involved with the smells, and he told the banker to "quit it" because he had heard voices telling him that the banker had drugs in the bank. Later he heard voices — including the voice of the man who had sold him his house — telling him to "Change your sewer system," which he proceeded to do. He had peculiar tactile sensations. He felt houseflies in an envelope, and the vibrations told him not to throw it away. Upon picking up the mail, he could tell through smell and touch whether or not to open it. He was aware of other smells on his children's school papers. He heard voices saying, "We don't care," and he thought this meant that he should kill himself. He almost did this with his shotgun but stopped after he thought about leaving his wife and children alone. There was evidence of thought broadcasting, thought insertion [the belief that someone else's thoughts were placed into his mind], and somatic hallucinations, including the feeling of a hammer hitting him on his tailbone, rump, scrotum, and penis. He said he had trouble maintaining an erection and ejaculated quickly. He said there were wires in his nose and head with repetitive banging, almost like a heartbeat.

At his wife's urging, he was hospitalized on two occasions in a general hospital near his home in November . . . and again in April [the next year], each time for about a 10-day period. After treatment with [trifluoperazine], he improved to some degree but about December . . . stopped taking his medication. In April [the next year] he was seen in the medical clinic of the university hospital, where he was treated for a minor skin rash. As his behavior was manifestly psychotic, he was admitted for the first time to the university hospital psychiatric service for a 3-week period. (This was his third hospital admission.)

On my teaching rounds I saw him shortly after his admission, at which time I also interviewed his wife. He responded favorably quite quickly to [haloperidol] and to hospital milieu treatment, and his delusions diminished but did not disappear. To deal with some of their marital conflicts, he and his wife began couple psychotherapy, but this could not be continued because they lived far from the hospital. He was referred to a psychiatrist in his community but failed to follow through. In July . . . on a return outpatient visit, he looked intent, shook one leg throughout much of the session, remained fixed in his delusion about the smells, and still believed that there was a tube in his penis. His wife corroborated the fact that he continued to act quite odd and at times had a fixed smile on his face. The patient, his wife, and usually the three youngest children were seen at 2-week intervals. Medication was changed to [thiothixene], which he tolerated better. When the smells came back, the patient became quite suspicious of his wife, accused her of extramarital sexual activity, and when driving his car, looked back constantly, believing someone was following him. As he became more disturbed, his wife reluctantly agreed to have him return to the hospital. This was his fourth admission. He remained with us for another 3 weeks, and during this time he felt he had a brain operation that had removed all of his senses, each of which was replaced intact except for his sense of smell.

In December . . . on his outpatient visit return, he reported that he had been taking his medicine regularly, said he was no longer troubled by smells, but was worried because he had fewer erections. However, he no longer believed that something from within his wife was put inside him. He blamed his wife for his lack of potency. Because he felt less like a man, he wondered if he should divorce her; yet he felt that he would miss her and the children. He was pursuing plans for a job through rehabilitation service. He still smiled somewhat vacuously, his wife looked more tired, and one sensed the extraordinary burden she was carrying with her six children and her sick, dependent, unpredictable, and at times frightening husband. The therapeutic program attempted to reinforce the patient's self-esteem; to clarify and at times to challenge his delusional notions; to insure his taking his medicine; and to encourage him to prepare himself for a wage-earning position. In addition, support was given to his wife through the county welfare social worker and through the psychiatrist who saw the patient.

The fifth hospital admission, in October [the next year], for a 2-week period, was the first on an involuntary basis. He became progressively less withdrawn and delusional after resuming his medication, and with

improvement, he was discharged with arrangements made again to introduce him into a vocational rehabilitation program. In December . . . the
patient started a job in automotive training 24 miles from his home. At
that time, he was better and the children were doing well. Early in
February [the next year] the wife called again to say that the patient was
becoming depressed and was not taking his medicine. She explained that
he worried about his maternal grandfather's health and felt he was
making both his grandparents worse.

The patient has had five hospital admissions since November 197[ ],
each for a period of less than 30 days. Although he lives some distance
from the hospital, he returns, albeit reluctantly, with his wife and three
youngest children for periodic visits. At the urging of his wife and the
psychiatrist, he does take the medication, but at times suddenly stops
taking it.

He remains unemployed, although efforts are repetitively made to
engage him in a rehabilitation program leading to a semiskilled job. He,
his wife, and six children are supported totally by county welfare funds.
The oldest son, age 10, because of past temper tantrums, was seen by a
school psychologist. The family is visited regularly by a county welfare
office social worker, who communicates with us about the state of the
family. The social worker and those of us in the hospital are continuously
concerned about the unpredictability of the patient and the possibility of
his injuring himself, the wife, or his children. The wife is hesitant to
initiate movement toward hospitalization because of his threats to kill
her if she persists in hospitalizing him.

The wife is clearly tyrannized by the patient's threats to harm her and
the children should she arrange for emergency hospitalization without
his consent. She is aware that she can call the police, who would respond
to her call and make a decision whether they believe the patient is
sufficiently sick or disturbed to be brought to the nearest county mental
health facility. Up to the present time the wife has been reluctant to do
this. She has also refused to initiate any involuntary hospitalization
through court order of the Town Justice. She has a good relationship
with the county welfare social worker and through her could also take
legal measures to arrange for involuntary detention of her husband.
(Romano, 1977, pp. 553–554)

Although we have been focusing on the description and definition of
schizophrenia in this chapter, the case history just presented permits the
opportunity to comment briefly on the need to view patients in their
family and work contexts and to consider the impact of schizophrenics on

their social networks. Romano asks, in the subheading preceding his description, "Freedom for Whom?" His question brings to mind F. Scott Fitzgerald's poignant phrase "the tyranny of the weak." Both remind us that concern should not end with the patients alone, but should embrace their families as well. Some analogue to an environmental impact statement should be a part of the discharge planning for schizophrenics. Attention to how the patient will affect his or her home and community environment and how the environment will affect the patient is equally important. There is a dynamic relationship between the two. Schizophrenics returned to stressful environments with demanding or belittling personal interactions do less well than those for whom after-hospital life is more tranquil, as we will see in Chapter 8. Although they are not sufficiently noticed, persons who must live with a schizophrenic need protection from unreasonable demands and even intimidation.

Schizophrenia's Forgeries and False Leads

In describing the phenomenology of schizophrenia, an early clinician used the term *intrapsychic ataxia* metaphorically to suggest a neurological defect, a loss of coordination of the mind. The phrase is vivid because schizophrenia's symptoms do suggest an imbalance and incoordination in the brain's circuitry, ranging from subtle to obvious. For the subtle aspects, Paul Meehl, a keen observer of schizophrenia for the past four decades, came up with the term *cognitive slippage*. In fact, LSD, (lysergic acid diethylamide), amphetamine (speed, crank, crystal, ice), phencyclidine (PCP, angel dust), cocaine (crack), and other street and "designer" drugs can produce short-duration symptoms that mimic genuine schizophrenia closely enough to lead to a false diagnosis. These and other drugs, such as bromide, carbon monoxide, alcohol, and digitalis, produce what can be called a drug-induced or *pharmacological phenocopy* of a schizophrenia-like psychosis. Head injury or disease such as a brain tumor, epilepsy, hypoglycemia, encephalitis, or uremia can produce some of the same schizophrenia-like symptoms. These could be called disease-induced or *somatic phenocopies*. Genetic disorders whose symptoms can be confused with schizophrenia include Huntington's disease, acute intermittent porphyria, and the sex chromosomal abnormalities XXY (Klinefelter's syndrome) and XXX females; for scientific writing and research, these are termed *genocopies*. The three groups of "forgeries" or imitators are summarized in Tables 2, 3, and 4.

TABLE 2

Selected drugs that can induce schizophrenia-like psychoses*

Alcohol (withdrawal hallucinosis)	Disulfiram (Antabuse)
Amantadine	Digitalis
Amphetamine (speed, crank, crystal, ice)	Ephedrine
	Ibuprofen (Motrin)
Atropine	Indomethacin
Bromide	Isoniazid
Bromocriptine	Levodopa
Carbon monoxide	Lidocain
Cannabis (marijuana)	LSD
Chloroquine	MAO inhibitors
Cimetidine	Methamphetamine
Clonidine	Pentazocine (Talwin)
Cocaine (and crack)	Phencyclidine (PCP, angel dust)
Corticosteroids (ACTH, cortisone, etc.)	Phenelzine (Nardil)
	Propanolol (Inderal)
Dexatrim	Propoxyphene (Darvon)
Diazepam (Valium)	Tricyclic antidepressants

*See the *Medical Letter*, August 17, 1984 and August 29, 1986 for other drugs.

Schizophrenia diagnosis requires delineating some length of time during which the patient has suffered symptoms. The DSM-III-R requires six months' duration, to avoid false diagnoses of schizophrenia for short-lived reactions to drugs or disasters (earthquakes, combat, plane crashes) that mimic many of the symptoms in genuine schizophrenia. The duration period is arbitrary and can lead to an underdiagnosis of genuine schizophrenia. The duration criterion gives rise to the use of another term, *schizophreniform psychosis*, for cases that resemble schizophrenia and that have lasted more than two weeks but less than six months. Europeans use the term schizophreniform psychosis more literally for cases in which the investigator does not want to exclude the possibility of true schizophrenia. In either instance, the term allows the diagnostician to red-flag an interim decision for later reconsideration. Everyone has a breaking point in the face of either acute or cumulative stress that may result in some kind of "psychogenic psychosis"; it is usually not schizophrenia and disappears after the stressors (or toxins) are removed.

TABLE 3

Possible somatic-disorder imitators of schizophrenia-like psychoses

Brain Injury or Disease	Metabolic and Systemic Diseases
Embolism (clot)	Vitamin B12 deficiency
Aqueduct stenosis	AIDS
Ischemia	Syphilis
Trauma (especially temporal lobe)	Tuberculous meningitis
Tumor (especially limbic and basal ganglion)	Pellagra (vitamin B6 deficiency)
	Hypoglycemia
Epilepsy (especially of the temporal lobe)	Hepatic (liver) encephalopathy
	Hyperthyroidism
Encephalitis (post-influenza, etc.)	Lead poisoning
Narcolepsy	Lupus erythematosus
	Multiple sclerosis
	Uremia
	Vasculitis

Schizophrenia is easily confused, particularly at first onset, with the other large category of severe "functional" mental illness, major affective disorders (depressive psychosis, mania, or manic-depressive illness), especially when bizarre delusions are present. The two illnesses appear at times so similar that misdiagnosis in both directions may contaminate some of the treatment and research literature. Mental health workers believe they are separate illnesses because they respond, by and large, to distinctive treatment, have different natural histories and outcomes, and have different family illness patterns. But until we can confirm a diagnosis chemically or by some genetic test, we can't be absolutely sure of the purity of our research data or treatment recommendations.

Experiments in Making Clinical Judgments

Careful observation of cross-national mental health statistics after World War II revealed a remarkable discrepancy between the frequencies of usage of the diagnoses of manic-depressive psychoses and of schizo-

TABLE 4

Possible genetic and chromosomal imitators of
schizophrenia-like psychoses*

XXY Klinefelter karyotype
XYY karyotype
XO Turner or Noonan syndrome
XXX karyotype
18q⁻ missing piece of long arm of chromosome 18
5,q11-q13 Triplication of this region of chromosome 5
Huntington's disease
Acute intermittent porphyria
Metachromatic leukodystrophy
Familial basal ganglia calcification
Homocystinuria
Phenylketonuria
Wilson's disease
Albinism
Congenital adrenal hyperplasia
G-6-PD deficiency (favism)
Kartegener's syndrome

*See McKusick (1990) for further information.

phrenia. British psychiatrists used the former diagnosis some 20 times
more often than did their U.S. counterparts, while the latter diagnosis
was used four times more often in the United States. Europeans not only
tended to use the stricter Kraepelinian diagnostic criteria for both categories, but also had a longer and more enthusiastic tradition for careful
classification. A 1972 study confirmed what researchers long suspected.
It compared 250 admissions to a Brooklyn, New York, state mental
hospital with the same number in a similar London hospital and found
that, on the hospital charts, nearly twice as many (65 percent) of the U.S.
patients were diagnosed schizophrenic compared to the U.K. patients. An
Anglo-American team of psychiatrists then used a structured interview
to collect basic mental health information on each patient in Brooklyn
and London. On the basis of such information, diagnoses on both sides of
the Atlantic converged dramatically. In the U.S. hospital, 32 percent of
the patients were determined to be schizophrenic; in the U.K. hospital,
26 percent were. Broadening the study to embrace other hospitals in New

York City and London produced similar results. In diagnoses based on structured, symptom-based information, 39 percent of the U.S. patients and 37 percent of the U.K. patients were called schizophrenic; 27 percent of U.S. and 24 percent of U.K. patients were called depressed, and 8 percent of U.S. and 7 percent of U.K. patients were called manic. A longitudinal picture of changing fashions in diagnostic practices between 1925 and 1970 in two world-renowned teaching hospitals in London and New York is shown in Figure 2 to illustrate the point.

A second major, multinational study, conducted by the World Health Organization, reveals additional important lessons about the diagnosis and classification of schizophrenia. It embraced 1,202 patients in nine countries: Colombia, Czechoslovakia, Denmark, Formosa (Taiwan), India, Nigeria, United Kingdom, United States, and USSR. Each patient received two diagnoses: a clinical one and a classification based on a computerized analysis of a structured interview. One diagnostic class was a nuclear syndrome that identified severely psychotic patients who suffered from a number of Schneider's first-rank symptoms listed in Table 1. This classification should have caught nearly every patient who was clinically diagnosed as schizophrenic. That it did not was both enlightening and unnerving. Almost all the patients whom the computer regarded as nuclear-syndrome schizophrenics were diagnosed schizophrenic by their examining physicians, but no less than 49 percent of those diag-

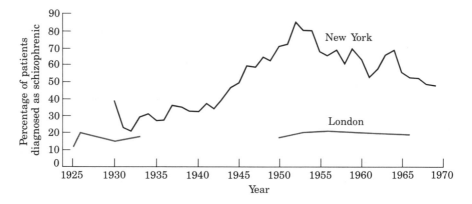

FIGURE 2.   The percentages of patients diagnosed as schizophrenic among all patients admitted to a United States Hospital — The New York State Psychiatric Institute in New York City — and a British Hospital — The Maudsley Hospital in London — revealing cross-national differences (U.S.-U.K.) and changes in "fashion" (U.S.) over the time period 1925 to 1970. Adapted from Kuriansky, Deming, and Gurland (1974).

nosed as schizophrenic by a physician were not classified by the computer as fitting into the nuclear syndrome. Even though a carefully structured interview was used to gather information, the computer missed many schizophrenics because it did not have a "whole-patient" mental health history. Its structured-interview information covered the patient's history for only the 30 days preceding the interview. When the computer was fed historical information, it agreed with the clinical evaluation in 90 percent of the cases. Without using a careful life-span perspective, diagnoses of schizophrenia are likely to be underreported. Accurate diagnosis is based on both present symptoms and past history.

A Prototype for Diagnosing Schizophrenia

If you feel frustrated by failing to find definitive answers to the central questions of this chapter—what is schizophrenia and who is a schizophrenic?—it is because the questions are very difficult to answer without qualifications. Let us try to reduce the ambiguity and possible frustration by defining a prototype for a diagnosis of schizophrenia. Figure 3 reveals, in detail, the proportion of all consensus-defined schizophrenics: those (306 of 811 clinically diagnosed cases) who each passed three hurdles—clinical, computer, and a statistical equation—and who had symptoms listed among the 27 symptom groups used by WHO to diagnose all varieties of mental illness. It is clear why one schizophrenic can be quite different clinically from another and why no one or two symptoms in isolation can infallibly pick out a genuine schizophrenic. For contrast, the symptom profile characterizing 99 WHO psychotic depressives (including manic-depressives in their depressed phase) is also shown in the figure. Although 97 percent of consensus schizophrenics showed "lack of insight," depressives were also characterized by this symptom at the high rate of 47 percent. The two symptom profiles highlight the great variation in symptoms experienced between patients in the two groups, as well as by patients within either prototype diagnosis.

The study showed, furthermore, that we are far from being able to predict outcome accurately from present diagnosis. None of the methods —clinical, computer, or statistical equation—were accurate predictors of future outcome. Some 27 percent of the study's schizophrenics experienced a single, short, psychotic episode and two years later had not relapsed. An equal percentage remained severely psychotic for the whole two years. No correlation between symptoms or history could predict any

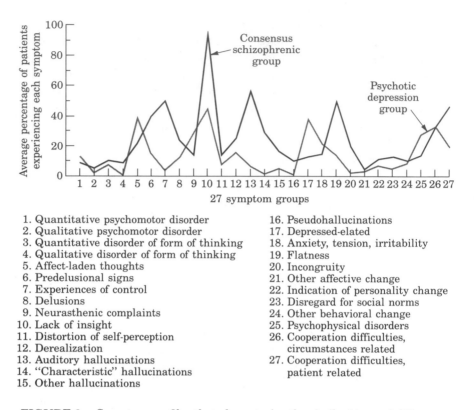

1. Quantitative psychomotor disorder
2. Qualitative psychomotor disorder
3. Quantitative disorder of form of thinking
4. Qualitative disorder of form of thinking
5. Affect-laden thoughts
6. Predelusional signs
7. Experiences of control
8. Delusions
9. Neurasthenic complaints
10. Lack of insight
11. Distortion of self-perception
12. Derealization
13. Auditory hallucinations
14. "Characteristic" hallucinations
15. Other hallucinations

16. Pseudohallucinations
17. Depressed-elated
18. Anxiety, tension, irritability
19. Flatness
20. Incongruity
21. Other affective change
22. Indication of personality change
23. Disregard for social norms
24. Other behavioral change
25. Psychophysical disorders
26. Cooperation difficulties, circumstances related
27. Cooperation difficulties, patient related

FIGURE 3. Symptom profiles that characterize the similarities and differences between consensus schizophrenics and psychotic depressives in World Health Organization cross-cultural research (1969–1971) and within the two diagnostic categories. Adapted from WHO (1973).

outcome. It may well be that, as a criterion for diagnosing schizophrenia retrospectively, outcome is a kind of fool's gold — a flaw in Kraepelin's emphasis on the course of the illness as a way to separate schizophrenia from manic-depressive disorders. Self-fulfilling prognostic statements about schizophrenics have hindered or delayed useful interventions — an unintended iatrogenic side effect. Too often, clinicians feel that the diagnosis of schizophrenia is equivalent to a death sentence and communicate that hopelessness to both the patients and their families. The accumulated data do not support such pessimism.

No matter what the purpose of schizophrenic diagnosis, whether it is that of a clinician faced with an individual patient and his or her family or that of a researcher seeking to gather and analyze family data, the great latitude of diagnostic criteria is extremely troubling. Diagnoses based on either too narrow *or* too wide sets of criteria give the least useful results, particularly for genetic study. Irving Gottesman and James Shields experimented elaborately with some of their data to prove this point. They established that fence-sitting between the extremes of too narrow and too wide is a useful and informative diagnostic posture.

They took 120 case histories originating from a study of schizophrenic twins at the Maudsley Hospital in London to six diagnosticians of varying persuasions, from strict to liberal, from Scandinavia, Japan, the United States, and the United Kingdom. They asked each judge to select the schizophrenics from these data. Then they pooled the results to obtain a six-person consensus. That gave them six individual opinions and one consensus opinion drawn from one data set. But that was not the end. Erik Essen-Möller, a senior Swedish psychiatrist with a strict diagnostic orientation, judged the cases; the cases were also classified by the Present State Examination (PSE) based criteria used by the WHO in the nine-country study noted above. Finally, a senior Danish clinician, Joseph Welner, reviewed the histories to give two further diagnoses: one based on traditional local definitions of schizophrenia (very Kraepelinian) and the other based on wider and more psychodynamic ideas developed from his role as the interviewer in the landmark Danish adoption studies led by David Rosenthal and Seymour Kety (see Chapter 7), which emphasized "spectrum disorders." The primary candidates for psychiatric conditions that may somehow be genetically related to schizophrenia are termed *spectrum disorders* and include schizotypal, schizoid, and paranoid personality disorders. The experiment, therefore, yielded 11 opinions for each twin. Under the narrowest interpretation of the WHO standard, only 17 were identified as nuclear syndrome schizophrenics with first-rank symptoms, quickly increasing to 52 schizophrenics with some relaxation of inclusion criteria among the 120 case histories; the Swedish psychiatrist identified 34 cases. The six-judge-panel consensus identified 69. The broadest diagnostic standards (used by Paul Meehl, the American clinical psychologist) identified 79. We will refer to these results again when it comes time to evaluate the differences in reported schizophrenia rates across cultures, across time, and across research studies. Notice already, however, the very wide spread: broad diagnostic criteria yield four and a half times as many schizophrenics as do narrow ones!

The experiment established the validity of middle-of-the-road criteria, as will be seen in Chapter 5, and illustrated both the problem of varied standards and the best solution. In the search for what kinds of disorders may be related genetically to schizophrenia, neither strict nor liberal diagnostic criteria served the search best. This is probably also true from a clinical viewpoint. If a clinician believes that only patients like the man described earlier (p. 24), who has been ill for months, should be called schizophrenic and treated as schizophrenics, many cases will be undiagnosed or misdiagnosed, and that fact may delay correct treatment. If, on the other hand, everyone who is mentally disturbed is labeled schizophrenic, clinicians are "crying wolf" and doing more harm than good to the patients by exposing them to inappropriate treatments. A narrow diagnostic standard does have a place in exploratory research on schizophrenia with the invasive techniques required by some of the very expensive, high-tech, brain-scanning equipment. Small samples of patients volunteer who must be "guaranteed" to be schizophrenic, and false positives cannot be tolerated.

Diagnosing Schizophrenia: An Integrated Summary

To summarize, let's describe the basic behavioral ingredients required for a clinical diagnosis of schizophrenia. We draw upon the accumulated clinical wisdom of Manfred Bleuler, the renowned Swiss psychiatrist whose father had introduced the term *schizophrenia* in 1908. Manfred Bleuler replaced his father at the Burghölzli Hospital in Zurich in 1942 and has studied and treated his patients intensively and extensively in an interactionist framework. For him, as for us, schizophrenia is a mental illness, a psychosis, a form of insanity, and not simply — as Ronald Laing, Thomas Szasz, or Herbert Marcuse and others have argued — the manifestation of a deviant or alternative lifestyle or evidence that society itself is sick. Individuals can be diagnosed as schizophrenic even if they are *currently in remission* and have few or no symptoms at the present time. This position is unfair to individual patients and could be misused against their civil rights, but it is necessary to advance the scientific quest for the origins of the disorder. Adults can also be diagnosed as schizophrenic even though they have experienced only one clear episode of a psychosis meeting accepted criteria, with relative normality or milder psychotic behavior before or after that episode.

Once other kinds of brain dysfunction such as those induced by trauma, tumor, toxins, or endocrine disease have been excluded, at least three of the following seven items would lead Bleuler to make a clinical diagnosis of schizophrenia:

1. An ordinary person would consider the patient's train of thought to be incomprehensible and obviously confused, i.e., ununderstandable.

2. The patient has a conspicuous incapacity for emotional empathy and this lack cannot be explained by any obvious fact.

3. The patient appears to be in a state of intensely abnormal excitement or stupor, which may last more than just a few days.

4. The patient experiences hallucinations and illusions that last more than a few days.

5. The patient experiences delusions.

6. The patient shows a sudden and total neglect of everyday, ordinary obligations or, without provocation, behaves with brutality toward family or strangers.

7. Friends and family members report that the person they know has suddenly become a *different* person, not like his or her old self, and the patient's behavior can no longer be understood.

We also believe, with Bleuler, that the appearance of purely manic or purely depressive phases before or after a clearly schizophrenic episode does not exclude the possibility of schizophrenia. We expect schizophrenics to have intermittent depressions — virtually nothing in their inner or outer lives brings them any pleasure — perhaps of a magnitude to warrant an additional diagnosis of a major depressive episode. This extra diagnosis should not obscure a schizophrenic diagnosis, but it should suggest that during the depressed period, precautions against suicide should be taken and antidepressant medications considered; at least 10 percent of schizophrenics commit suicide, a rate 10 times greater than that for males and 18 times greater than that for females in the general population.

Schizophrenia and Childhood Psychoses

Before continuing our search for the causes of schizophrenia, it is necessary to clarify the relation of childhood psychoses, including childhood schizophrenia and infantile autism, to schizophrenia as it has been described in this chapter. Is such information relevant, or can it be safely excluded from further consideration as evidence? Scientists have speculated for years about the possible connection between infantile autism (and other early childhood psychoses) and adult schizophrenia, some instances of which begin at puberty. Are these severe childhood illnesses, termed *pervasive developmental disorders* in the DSM, simply early variations of adult schizophrenia, or should they be considered separate conditions, distinct not only from schizophrenia but also from each other? The difficulty in answering this question lies in part in the difficulty of rendering a psychiatric diagnosis for disorders of development and disorders of social interaction in young children. One can listen and watch, but for the most part, young children and infants can't talk about how or what they think, see, or feel. Childhood psychoses have not been as thoroughly studied as the adult conditions we have already described. Some recent attempts to define criteria sets for diagnosing deviations from normal development, as well as the move by researchers from case studies to population-based research designs, will improve the state of knowledge. We join with a veteran clinician and researcher, Lorna Wing of London, in the opinion that the terms *childhood psychosis, childhood (infantile) autism,* and *childhood schizophrenia* have seen their day and require replacement by more precise terms.

Psychoses that begin in infancy and early childhood differ in important ways from psychoses that begin at and around the onset of puberty. The distinctions are both behavioral and genetic and suggest the existence of disorders that are not the same as schizophrenia. Life would be simpler if we could lump all such deviations together, but the accumulated data do not support doing so.

The clinical picture observed in early-onset childhood psychoses, including classical *early infantile autism* described by the child psychiatrist Leo Kanner in 1943, closely resembles what is seen in children with known brain damage who are mentally retarded and behaviorally disturbed. About half of all autistic children are severely retarded, with IQs under 50; schizophrenics have a more or less normal distribution for IQ. Extreme social aloofness and indifference, extending to parents, siblings, or strangers, and elaborate, stereotyped rituals frequently characterize the behavior of these children, but are seldom seen in schizophrenics. Hallucinations and delusions, hallmarks of schizophrenia, are rare or absent in early-onset psychotics followed up into adulthood. Another markedly different feature between early-onset psychoses and the schizophrenics is the course of illness; only in the latter do we observe episodes of remissions and then relapses. For childhood psychoses that begin around the time of puberty, however, the clinical picture is often indistinguishable from that of the schizophrenias described in the first part of this chapter.

There are also some epidemiological patterns that support the belief that early-onset psychoses are different from schizophrenia. The 1:1 sex ratio seen in schizophrenia can be contrasted with the male preponderance in childhood psychoses, which ranges from 2:1 to 5:1 in different studies. Compared to the lifetime prevalence of 10 in 1,000 adults having schizophrenia, the pervasive developmental disorders are quite rare. Depending on the narrowness of the definition, the prevalence of these disorders in the population under age 15 ranges from 1 in 10,000 to 20 in 10,000. Much has been written about these conditions that has led to the incorrect impression that they are as common as schizophrenia. Since mental retardation from various causes affects some 1 to 2 percent of the population, and since the symptoms overlap with the "childhood psychoses," diagnostic confusion and overdiagnosis of the latter are inevitable.

Schizophrenia, as noted earlier, tends to run in families. With a couple of exceptions, studies of the parents of early-onset psychotics show nearly a zero rate of schizophrenia or other psychoses; the parents of pubertal-onset cases, however, have a much higher rate, 9 percent, even higher than in parents of adult-onset schizophrenics. A similar pattern appears in studies of siblings: Only some 2 percent of the siblings of early-onset psychoses, contrasted with a minimum of 7 percent of the siblings of those with onsets after age 5, suffer from psychoses of any kind. Based on currently available data, we tentatively conclude that pubertal-onset psychoses are rare and severe early-onset instances of schizophrenia and that pervasive developmental disorders beginning before age 7 or so are qualitatively different from schizophrenia. These conclusions, however, should be viewed as challenges to research rather than as established facts, because knowledge about pervasive developmental disorders is still in its scientific infancy.

The next chapter will introduce first-person accounts of schizophrenics themselves to add flesh to the scientific bones describing, defining, and diagnosing schizophrenia so far. That will be followed by a series of chapters detailing the fascinating pursuit of clues to the major kinds of evidence gathered from different vantage points that inform our current understanding of the genesis of schizophrenia.

3

Anguished Voices I
Personal Accounts of Survivors

An authentic, multidimensional representation of schizophrenia and what its presence does to patients and their families cannot be readily drawn from the piecemeal-scientific approaches to diagnosis and symptom description presented in the last chapter. Oral histories, or first-person accounts, have served our colleagues in the humanities well in bringing history to life. We hope to accomplish the same kind of enlivening by presenting the voluntary life stories (not formal life or case histories) contributed by some sensitive and articulate present or former schizophrenic patients, in their own words. (In Chapter 9, family members relate their personal experiences with someone afflicted with schizophrenia.) The words were written, not tape recorded, and thus take on a more orderly form than would the typical oral history. Very often, such orienting information as age at onset, developmental history, and medication status are not given.

The accounts were written in response to a standing invitation from the editors of the *Schizophrenia Bulletin*, a professional quarterly journal published by the U.S. Superintendent of Documents, Government Printing Office, in Washington, D.C., for the Alcohol, Drug Abuse, and Mental Health Administration (ADAMHA) of the Public Health Service. The contributions are in the public domain; the authors are identified in footnotes as Anonymous or by a pseudonym or their real name, depending on each author's preference. We have held our comments and clarifications to a minimum, and they are given within square brackets. We invite you to listen to these voices nonjudgmentally, and empathetically.

Where Did I Go?
An artist describes her one-year experience of schizophrenia.[1]

The reflection in the store window — it's me, isn't it? I know it is, but it's hard to tell. Glassy shadows, polished pastels, a jigsaw puzzle of my body, face, and clothes, with pieces disappearing whenever I move. And, if I want to reach out to touch me, I feel nothing but a slippery coldness. Yet I sense that it's me. I just know.

I know I'm a 37-year-old woman, a sculptor, a writer, a worker. I live alone. I know all of this, but, like the reflection in the glass, my existence seems undefined — more a mirage that I keep reaching for, but never can touch.

I've been feeling this way for almost a year now, ever since I was diagnosed a paranoid schizophrenic. Sometimes, though, I wonder if I ever knew myself, or merely played the parts that were acceptable, just so that I could fit in somewhere. But the illness has certainly stripped me of any pretense now, leaving me, instead, feeling hollow, yet hurting. I twist and turn, hoping to find a comfortable position in which to be just me.

There are still occasional episodes of hallucinations, delusions, and terrible fears, and I have medication for these times. It relieves my mental stress, but I hate my bodily responses to it and the dulling of my healthy emotions. Therefore, I stop using the drug as soon as the storms in my mind subside. And I keep wondering why there isn't more emphasis on alternative therapies, such as the holistic programs used now by people with physical illnesses.

---

[1] M. E. McGrath, *Schizophrenia Bulletin*, 1984, 10: 638–640.

So I've searched, in library books and in articles about schizophrenia, hoping to find other solutions and answers to my whys, how longs, what's the cure. Some of the information is frightening—the case histories of patients, the descriptions of symptoms. Some of it is confusing, reaming with speculations, yet with every author being certain that his written word is better than the last answer in print. Schizophrenia is genetic— no, no, it's surely biochemical—definitely nutritional—sorry, but it's caused by family interactions, maybe stress, etc. Now, with the worship of the technological gods, the explanation is that schizophrenia is a brain disease colorfully mapped out by the PET scanner. I suddenly feel that my humanity has been sacrificed to a computer printout, that the researchers have dissected me without realizing that I'm still alive. I'm not comfortable or safe in all their certain uncertainties—I feel they're losing me, the person, more and more.

In the most recently published book I've read, a doctor writes that psychotherapy is useless with schizophrenics. How could he even suggest that without knowing me, the one over here in this corner, who finds a lot of support, understanding, and acceptance with my therapists? Marianne is not afraid to travel with me in my fearful times. She listens when I need to release some of the "poisons" in my mind. She offers advice when I'm having difficulty with just daily living. She sees me as a human being and not only a body to shovel pills into or a cerebral mass in some laboratory. Psychotherapy is important to me, and it does help.

I sound angry—I guess I am—at the illness for invading my life and making me feel so unsure of myself . . . at the medical researchers who now only want to pick and probe into brains or wherever so they can program measurements into their computers while ignoring me, the person . . . at all the literature which shrouds schizophrenia in negativity, making any experience connected with it crazy and unacceptable . . . at the pharmaceutical industry for being satisfied that their pills keep me "functional" when all the while I feel drugged and unreal to myself. And I'm angry at me for believing and trusting too much in all this information and becoming nothing more than a patient, a victim of some intangible illness. It's no wonder to me anymore why I feel I've lost myself, why my existence seems a waning reflection.

But I'm still searching, questioning—looking inside now rather than on the library shelves—just wanting to feel a little comfortable. I know all the negatives: schizophrenia is painful, and it is craziness when I hear voices, when I believe that people are following me, wanting to snatch my very soul. I am frightened too when every whisper, every laugh is about me; when newspapers suddenly contain cures, four-letter words shouting at me; when sparkles of light are demon eyes. Schizophrenia is frustrating when I can't hold onto thoughts; when conversation is projected on my mind but won't come out of my mouth; when I can't write sentences but only senseless rhymes; when my eyes and ears drown in a flood of sights and sounds . . . and on and on, always more. . . .

But I know I'm still me in the experience. And I'm creative, sensitive. I believe in mysteries, magic, rainbows, and full moons. I wonder why it's expected that I be quieted, medicated whenever it seems I'm stepping out of the boundaries of "reality." Should I let anyone know that there are moments, just moments, in the schizophrenia that are "special"? When I feel that I'm traveling to someplace I can't go to "normally"? Where there's an awareness, a different sort of vision allowed me? Moments which I can't make myself believe are just symptoms of craziness and nothing more.

What's so "special"? Well, the times when colors appear brighter, alluring almost, and my attention is drawn into the shadows, the lights, the intricate patterns of textures, the bold outlines of objects around me. It's as if all things have more of an existence that I do, that I've gone around the corner of humanity to witness another world where my seeing, hearing, and touching are intensified, and everything is a wonder.

Music, especially if I listen through headphones, envelopes me and becomes alive, breathing high and low notes, and I'm floating on the movement.

Sometimes, in my schizophrenia, I go to the library, feeling like an explorer in a jungle of words and pictures. It can be frustrating because I capture nothing — not even one book chosen and checked out — but I scan the photos, the copied art works, even focus on a paragraph or two, as I venture along the shelves, my eyes jumping from book to book. I soon leave, empty handed, yet satisfied by having seen so much.

My illness is a journey of fear, often paralyzing, mostly painful. If only someone could put a Bandaid on the wound . . . but where? Sometimes I feel I can't stand it any longer. It hurts too much, and I'm desperate to feel safe, comforted. It seems, at these times, when I reach bottom, that I'm given a message and I feel mystical, spiritual, and like a prophet who must tell anyone that there's really nothing to fear. A white light often appears, branding this message on my very soul, and those who are most afraid will see it in me and be at peace. And I somehow feel better for being the courier.

These "special" moments of mine — there are so few, but I look for them and use them to help me pass through the schizophrenic episodes. And I can't even predict when or if these moments will come. But I won't deny their existence; I won't tell myself it's all craziness.

I'm hopeful about the ongoing research to find an answer to schizo-phrenia, and I'm grateful for all the caring and the help of those in the mental health profession. But I know that I'm the schizophrenic living the experience, and I must look inside myself also for some ways to handle it. I have to be able to see me again as a real person and not a fading reflection.

The Messiah Quest

A young schizophrenic describes his vivid delusions and hallucinations, writing in the third person to distance himself from the episode. Brief comments by two therapists who saw the author during different phases of the experiences described are appended to this account.[2]

"In spite of certain passages in the text, I had never previously felt fervently about any religion, religious figure, or telepathy, and I had never felt harassed by the Central Intelligence Agency. The experience described followed many extreme stresses — death of a parent, end of a longtime romantic relationship, and a career change. Before these events, my emotional and social adjustment had been good . . ."

A drama that profoundly transformed David Zelt began at a conference on human psychology. David respected the speakers as scholars and wanted their approval of a paper he had written about telepathy. A week before the conference, David had sent his paper "On the Origins of Telepathy" to one speaker, and the other speakers had all read it. He proposed the novel scientific idea that telepathy could only be optimally studied during the process of birth. He believed that the mother and infant have a telepathic bond that begins during delivery and should be studied before stimuli in the outer world significantly influence it. The paper described his observation, in an obstetrics clinic, of the mother and infant's facial expressions. They smiled or cried in parallel during delivery and for several minutes afterward. Facial expressions of happiness or pain appeared to occur at the same time and to be of similar intensity. He hoped this correlate, consistently present at seven births, would be verified for all humans. David knew that the paper, in reflecting engagement

---

[2]David Zelt (pseudonym), *Schizophrenia Bulletin*, 1981, 7:527–531.

with an esoteric subject, was a signpost of his growing retreat from mundane reality.

David's paper was viewed as a monumental contribution to the conference and potentially to psychology in general. If scientifically verified, his concept of telepathy, universally present at birth and measurable, might have as much influence as the basic ideas of Darwin and Freud.

Each speaker focused on David. By using allusions and nonverbal communications that included pointing and glancing, each illuminated different aspects of David's contribution. Although his name was never mentioned, the speakers enticed David into feeling that he had accomplished something supernatural in writing the paper.

A spiritually evolved person with great capabilities was the center of attention. Extraordinary powers of perception, a gift for telepathy, and the intellectual prowess of an Einstein were mentioned. David was certain that all of these allusions were to him when one speaker, while discussing the telepathy hypothesis, said, "Our shepherd." He was compared to a lion — courageous, regal, and wholesome; or a bird that could soar high like an eagle — extremely intuitive. He felt glorified.

David was described as having a halo around his head, and the Second Coming was announced as forthcoming. Messianic feelings took hold of him. His mission would be to aid the poor and needy, especially in underdeveloped countries. He also wanted to help everyone appreciate the joys and bear the sorrows of life; he hoped that, partly from his efforts, people would become more sensitive, caring, understanding, and loving of others.

David's sensitivity to nonverbal communication was extreme; he was adept at reading people's minds. His perceptual powers were so developed that he could not discriminate between telepathic reception and spoken language by others. He was distracted by others in a way that he had never been before. It was if the nonverbal behavior of people interacting with him was a kind of a code. Facial expressions, gestures, and postures of others often determined what he felt and thought.

Several hundred people at the conference were talking about David. He was the subject of enormous mystery, profound in his silence. Criticism, though, was often expressed by skeptics of the anticipated Second Coming. David felt the intense communication about him as torturous. He wished the talking, nonverbal behavior, and pervasive train of thoughts about him would stop.

One speaker, whose name was a household word, referred to this intense communication about David as Nazism. The denouncement had a dramatic effect: People then sat passively and without talking or thinking. They no longer communicated to David through movements, speech, or telepathy. Revering this speaker and in an extremely suggestible state, David felt that verbal and nonverbal harassment of anyone very spiritually evolved was like Nazism. He hated Nazism, the worst of all evils in his view. For the first time, David anticipated that his role in life would be to see that the psychological brutality of Nazism could never happen to anyone else. He would ensure that humanity would be cleansed of the potential for Nazism. This transformation of humanity would take place with the arrival of the Second Coming.

During the next few weeks, David came to believe that he was the reborn figure of Jesus Christ and that their spirits were identical. Like Christ, he was constantly in touch with the infinite and the eternal, and lived with a halo around his head that represented unity with God. David believed that he was the only person who could prevent the impending war that would end the world. He would prevent it by loving all humans and never qualifying or compromising his love.

From his apartment, David had a panoramic view in which many people expressed the basic needs and goals of humanity to him. They hoped that as the Messiah he could promote world change for the better. Activity on the left side represented the transcendental, intuitive, and holistic realm of human life; activity on the right side represented the material, rational, and analytical realm. As an example of expressed needs and goals, exhaust from airplanes in the sky formed patterns: the balanced scales of Libra on the left symbolized emotional harmony, and a scorpion on the right symbolized the analytical approach to life. . . .

David began to suspect and then perceive that a federal agency was observing him. For a moment of insight explaining many peculiar, recent events in his life, he knew that he had been accused of treason for slandering Americans during his psychotherapy. Specifically, he had told his psychiatrist that Americans were debasing him "like Nazis." Wondering whether the Federal Bureau of Investigation was observing him, David decided to employ his telepathic powers to find out. An agent at a local agency told him there was no investigation. David read the agent's mind, however, and determined that the Central Intelligence Agency was conducting an investigation.

By electronic means, the CIA let people around the world know what David said in therapy. Gradually, everyone took sides either for or against the CIA and their opponent, David. The speakers from the conference all promoted David in their own countries as the Messiah. To harass him, verbally or nonverbally, was the psychological equivalent of Nazi torture. Many countries were bitter toward the United States government for letting the CIA investigate a spiritually evolved person.

David viewed the CIA, in its harassment, as violating an amendment of the Constitution — freedom of religion.

It dawned upon David that the CIA was listening to most of his thoughts wherever he went, even sometimes during sleep. David could not think privately in words. His thoughts in words gave rise to subvocal movements that produced specific patterns of sounds during breathing; the patterns were immediately picked up and deciphered by hidden CIA electronic equipment. David had no consistent privacy of mind, except for concrete visual imagery; such imagery was usually very pleasurable and could not be monitored because no sounds were generated. Thoughts in words often seem to come from external sources that were localized in space, as if someone were talking to David; he could hear these thoughts, simultaneously while thinking them or momentarily later, broadcast by the CIA using electronic means. Wanting a confession of treason, the CIA tormented David by playing his thoughts aloud and also by making comments and criticisms about his thoughts. The CIA always claimed only to want the truth of whether David was a "traitor" in his own mind; nonetheless they tormented him. The CIA treatment of him made the flow of his thinking sometimes seem chaotic or random. Also, playing his thoughts aloud led to a vicious cycle that was extremely painful for David. Hearing a particular thought broadcast tended to promote repetition of that thought.

Because his thoughts were broadcast around him, David often felt that his consciousness was controlled from outside himself and that he had merged with the external environment. In his perceptions, broadcast thoughts were often superimposed on the sounds of running water, air vents, and cars passing by. These continuous noises of water, air, or car motion usually sounded like flies buzzing in the background of the broadcast thought. The boundary between his mind and the world seem blurred. . . .

David's thought processes and communication with others occurred in two basic ways: One was adapted to the rational realities of others, and the other way — the code — was magical, poetical, and fantastical. The code was used in his continuous struggles with the CIA and sometimes in communication with God. As time went on, the code came to dominate the functioning of his psyche. David never told anyone about the code, which first entered his awareness at the conference. In his perceptions, feelings, and thoughts, the code gradually influenced everything except for the use of words per se — influenced telepathy, facial expressions, gestures, postures, and intensity, pitch, rate, rhythm, and pauses of speech. His psyche generally seemed in an altered state to others because communication except that with God, was often difficult to do in a logical way; communication with God, however, was sometimes experienced by David according to the code. . . .

Whenever David had contact with the media, the code was used. For example, the television and radio stations used electrical signals and verbal messages to convey their attitudes about David. The electrical signals, such as a flicker on the television set or a burst of static on the radio, meant that the immediately preceding remarks referred to David. The verbal messages were usually either praiseworthy or condemnatory, and were understood in some way by everyone who was watching television or listening to the radio during these moments. The stations would state terms such as Jesus Christ or schizophrenia and then wait for David's response in thoughts, which would be broadcast by the hidden CIA equipment. On television, NBC had a transcendental outlook and often designated David as Jesus Christ. CBS usually described David as having schizophrenia. ABC expressed mixed feelings; either David was divine or ill. Sometimes, these television stations referred to David with advertisements, each of which portrayed a specific message about him. As examples of messages, Paul Newman thought that David was exceptionally gifted at telepathy and an endangered species like the American eagle; Merv Griffin said, "We love you," because David had brought world attention to the harm that the CIA was doing to the world; and Jean Stapleton, alluding to CIA slandering of David, said, "A thousand words don't tell a picture." Generally though the degree of attention was so painful that David tried to avoid contact with any media. . . .

Everyday, David studied the patterns in the sky formed from clouds of airplane exhaust. These patterns, always vivid, typically expressed a favorable view of David. Commenting on the war between David and the CIA, people often said, "All the skies are on your side." The wings of an angel, a lion, an eagle, and artist's pastels were commonly presented in the sky. David also sometimes saw a lion with two heads, a lion with wings, or an eagle with the head separated from the body. . . .

Several months after the conference, David's vision of emotional change among humanity had become clarified. God suggested that eventually humanity, guided by David, would unite in love. In this global state of love, there would be no need for the negative emotions of anger and disgust. To David, anger and disgust represented failures at empathy. He fully believed that lasting, empathic atonement by all humans with one another was the sole way to prevent the end of the world. He hoped that a new unity among mankind would arise from the worldwide division over the CIA conflict.

David knew that whatever was wrong with the world lay within the psyche. As [he was] the Messiah, his psyche mirrored humanity's problems. He struggled against profound powers — Nazism, the CIA, and the self-destructive tendencies of humanity — but he knew that the power of God would ultimately triumph. David's duty was to fulfill God's ultimate goal — to turn the Earth into a heavenly kingdom. He wanted to do everything he could to reach this goal. In heart and mind, David gave himself fully to his Messiah quest.

COMMENTS FROM DAVID'S THERAPISTS

FIRST THERAPIST. I was David's therapist during his acute episode, and he and I decided to go through it without medication. This was his intention and request, and my preference as well. It is a hard course to take because fear and suffering are all too evident. However, David's outcome provides the final justification for such a therapeutic policy. His account takes the reader directly into the feeling of the intense experience of psychotic inflation and the accompanying fears. My part was mainly to empathize and to remind him often that the phenomena belonged to his inner world. It was impressive to me to find someone with an exhaustive intellectual knowledge about psychosis still unable to bring his critical faculties to bear upon the onslaught of ideation. This ideation was visionary in nature and scope, although overwhelming. In time, his critical faculties did work effectively, and one must acknowledge that the outcome has been favorable. The fact that he is well endowed with high intelligence has, I think, played a vital part in his capacity to steer his episode toward growth and development. What strikes me most about his present state is his heartiness, vigor, and warmth; he seems to have abundant energy and high motivation, and to relate with fullness of feeling.

SECOND THERAPIST. I felt that being available to David on a daily basis was essential for him and that not doing so would have been cruel. There were few people with whom he could share his feelings and ideas. Our meetings, an open exchange between two people, seemed to ameliorate the loneliness of his struggle and perhaps were helpful in the resolution of his struggle. Despite intense involvement with his inner life, he was able to function in the world on a simplified level. Besides taking care of himself, he went out and did things with people and also read. His writing is a demonstration of the usefulness of the experience and reflects his change for the wiser.

After the Funny Farm

A mental health worker recollects her bittersweet, mainly bitter, inpatient experiences and calls for reforms in certain dehumanizing practices.[3]

We had met under the most unusual circumstances, in a place we came to call "The Funny Farm."

Shortly before our meeting, I had seen the movie *One Flew Over the Cuckoo's Nest*. I enjoyed the movie [adapted from the book of the same title by Ken Kesey] as an art form, for achieving the delicate balance between comedy and tragedy that exists in all of our lives. I recall laughing at the antics of the patients and silently applauding their ability to laugh in a world reserved for sadness and silent suffering.

Later, as a cuckoo in the nest, I found the experience of hospitalization much less amusing. Medication upon medication made my thoughts return to reality, but my body seemed suspended in time and space. There was no laughter; there were no tears; there was only existence. The poignantly painful ending of the movie reappeared as my own personal pain. I, too, had been symbolically suffocated in the name of caring and love. The major difference between me and the movie character was that I had survived.

At the sound of "Say 'You Love Me'" by Fleetwood Mac, Penny danced into my room and my life. We were a study in contrasts: My long, dark hair framed a face adorned with sweet innocence. Penny's short blonde hair outlined a round face carefully painted and subtly pretty. My only jewelry was a tarnished gold cross, which hung around my long, thin neck. My frail body carried a yellow nightgown like a shroud. Penny's fingers and wrists were wrapped with silver and gold, bearing chunks of turquoise.

---

[3]Anonymous, *Schizophrenia Bulletin*, 1980, 6: 544–546. After careful consideration, I have decided to publish this article anonymously, in the hope that by doing so, I will protect my family, my friends, and myself from any further embarrassment and discrimination. Protection against stigmatization is needed because our society does not feel "safe" for those of us who have been hospitalized for mental illness. As a former patient and employee of the National Institute of Mental Health, I am hopeful that this article will serve as a catalyst for needed change.

As Penny danced rhythmically from room to room, I tiptoed angelically in this strange, new world. Her laughter and mania blanketed her depression. My silence shielded my private conversations with God, who directed me in ways I could not understand, but [I] never failed to obey. Penny spoke of oil wells and the Kansas City Mafia. I spoke of the Scriptures, Easter, faith, forgiveness, and love.

Penny brought laughter and happiness. She was generous — willing me an oil well — in hopes of lifting the darkness and relieving the pain. She came into my life a personal Santa Claus.

We harmed no one in that place, except perhaps ourselves, but much harm was done to us there. We were initiated into a stigmatizing sorority. We shall never forget the horrors of drug lines; group meetings (called milieu therapy) where strangers shared personal problems and revealed intimate details of the world beyond that place; workshops (called occupational therapy) where patient staff persons (whose commands sometimes bubbled with anger) tried to teach tranquilized patients to paint ceramic mugs and ashtrays; recreation (called physical therapy) where exercise was trying to touch your own toes, arranging the marbles on the Chinese checkerboard, and chasing ping-pong balls.

There is pain just remembering, but there is also humor. It's funny remembering a place called a hospital that is more like a farm — a place where nurses serve as ranch hands, placing feed at regular intervals in tiny paper cups across Dutch doors; where unpredictable human behavior is viewed as "crazy" and where predictable animal behavior is viewed as ideal; where doctors seem more like farmers, using tranquilizers as cattle prods; where the day is spent grazing on the pasture of insanity — doing nothing useful or harmful while there, preparing to go into the world as a fat calf to be society's scapegoat.

When we (Penny and I, and my mom before me) had learned our lessons well, we were released on good behavior to the slaughterhouse of society.

We had been trained well over the years, even before going to the funny farm, to be diligent workhorses. Even now we serve our masters loyally and indefatigably. Our rewards are shelter and food — the basics. There are a few who love us; some who respect our courage; some who do not understand and fear us. We typically move quietly in the world asking for little, expecting the worst, and longing for human dignity and equality.

For 2 years after leaving the funny farm, Penny and I moved in our separate worlds, making infrequent contact with each other. Once we began to laugh a little at ourselves and the funny farm, we scheduled more frequent get-togethers. Now we meet weekly. In exchange for sewing lessons, Penny teaches me to laugh. She asks me to repeat stories of funny farm happenings, as though they were favorite fairy tales. We both find great pleasure in recalling the time when I, as God's obedient servant, ran away from the hospital in my nightgown, conned two unsuspecting gardeners out of their pickup truck, and returned like a recaptured prisoner under the escort of three hospital "heavies." We often retrace in our minds the scene of walking around and around the hospital building — a privilege granted to those of us who had progressed to the level of trustworthiness. Upon recalling a return trip to visit a hospitalized friend, Penny remarked to me in jest, "Would you believe that they [the patients] are still walking around in that place?"

Generally, our humor is private and personalized, always harmless to others. My mom laughs at dirty jokes, never at the actions of others, no matter how bizarre they may appear. Penny turns to typing as an outlet, remembering with bellows of laughter the therapeutic breakthrough of typing four-letter words repeatedly for hours. To bring humor into a depressing day, I retreat to a restroom stall set aside for the handicapped where I smile in memory of the time I used the men's room in the corridor outside my psychiatrist's office, secured myself safely from my private witch doctor, and sang "Holy, Holy, Holy" as loudly as I possibly could until he came to claim me.

Mental handicap creates new "stalls" for those traumatized by hospitalization for mental illness, just as it does for those who are crippled physically. Penny, my mom, and I (all three of us) have experienced the problems and barriers that lie before us in "normal" society. My mom could not cope with the cruelty of a world that imagined her traveling by broom and making dinner for us in a cauldron over an open flame. She retired on disability.

Penny blew the union whistle when her bosses tried to send her to the glue factory of disability retirement. Overnight she was offered a job in a place previously and repeatedly described as closed: No vacancies. She characterizes her new job as a bureaucratic pacifier with little, if any, advancement opportunity.

For me, the scene has been repeated in many different settings: a supervisor who viewed my work and abilities as outstanding and my rate of productivity as very high before my illness, but who recommended disability retirement when I was depressed and less productive; a university that graduated me with high honors, admitted me into its graduate program with outstanding recommendations, and then sent me a form letter in response to my request for readmission (following my illness) saying, "You do not meet our admission requirements"; and community mental health agencies that rejected my offers to be of assistance because I "scared" mental health professionals. The successful outcomes realized since then have been costly in time and energy: With the aid of an attorney (who has become the best kind of therapist), I am now in a job that offers hope for a brighter future; a Master's degree with high honors was recently conferred upon me; and I have my choice of affiliations with community service organizations.

The literature says little about us individually. Most researchers group us, thereby reinforcing the stigma. Some lay odds on our recovery and predict high rates of suicide. Some experiment with us, offering convincing evidence that we can be trained — rehabilitated. Others raise ethical concerns about studying us, but justify their actions by noting that useful data can be obtained by following us. Some have tried to document that public attitudes toward the mentally ill have changed.

If my own research and experiences are representative, public attitudes have not changed. From my perspective, researchers continue to define stigma with statistics. Physicians continue to locate emotional pain points with questions. Families continue to treat mental illness as a silent, shameful disease. Clergymen continue to preach that mental illness is the result of satanic influence. The barriers remain. They are real.

I am glad now that I went to the funny farm. I consider it the very best training I could have had as a mental health professional. I know firsthand what it is to be fully clothed and feel stripped naked. I know what it means to be labeled "mentally ill," "handicapped," "schizophrenic," "multiple personality," "manic-depressive"; to be assigned a diagnostic code with decimal points for clarification; to file claims for reimbursement of medical services detailing mental problems, as opposed to generalizing physical ailments. I don't need to read a textbook to understand the meaning of psychosis and neurosis. I know how it feels to be a guinea pig; to shuffle under the influence of Haldol; to sleep under the influence of Dalmane; to lose my hair under the influence of Lithium. I know the joy of insanity and the hell of an insane asylum. I know about those who call themselves doctors, caregivers, and clergymen, and who sit in judgment of those whose behavior and thoughts they do not understand.

I don't need to read about snake pits or rose gardens or cuckoo's nests or minds that found themselves or *The Inner World of Mental Illness* [B. Kaplan, Ed., 1964] even though I have read them. I've written my own chapters in the book of life. Fortunately, I have a mother who instilled in me the spirit of our Indian ancestors, and I have Penny who reminds me of the funny farm and makes me laugh at the greatest tragedy of my life.

The Hard Road Back

Acute and painful observations of a pharmacy student as both a patient and a professional, and her informed response to drugs that she is taking and learning to dispense.[4]

On my first day of externship at a hospital I was waiting for the pharmacy to open since I was early by an hour. I was sitting in a lobby wearing a white lab coat and required name tag and catching a catnap. A young male patient approached me.

"Excuse me, miss. Do you know that every morning when I get out of bed I feel there is danger everywhere?"

Because I had a white coat, he assumed I had an answer to this problem, or at least that I was not in his situation and could offer some assistance. I was taken off guard by this psychiatric patient but said, "You sound very frightened." He said, "Yes. Are you sad or just resting?" I felt he was seeing right through me. "No," I replied, "I'm not sad; I'm just very tired." "Oh," he said, "then I'll leave you alone," and he walked away.

My white coat and name tag offered me no immunity from schizophrenia. Pharmacy students are vulnerable just as everyone else is, in spite of the fact that we are taught about all diseases as if we were an immune group.

Inside, while I spoke to this patient I wanted to say: "Yes, I too sense danger everywhere, each morning and all day. It's hard for me to get out of bed, to go out of the house, to talk to people; it's hard just to get dressed and get outside and function. I'm afraid of people, of change. I'm sensitive to sunlight and noise. I never watch the news or read a newspaper because it frightens me."

Talking to this patient had made the conflict within me very obvious. This young man and I are related in a way I cannot share with him, with my fellow students, or with faculty. Yes, I am a pharmacy student. But, yes, I have been diagnosed as schizophrenic and have been hospitalized on three occasions when I could not function. Yes, I am on neuroleptics and must see a psychologist at least once a week, and sometimes more often, in order to function.

I wanted to tell him, "Yes, I know how it feels, and isn't it terrible?" Realization of this fact makes my role as a pharmacy student seem

[4]Anonymous, *Schizophrenia Bulletin*, 1983, 9: 152–155.

artificial — almost as if I must pretend and cover up to get by and pass as "normal" — and then there is always that danger that under stress or pressure my schizophrenia will get out of control and I will be found out.

In lectures on antipsychotic drugs I want to tell faculty and fellow students what it feels like to take these medicines and have to depend on them to function "outside" and what it is like to be titrated as an individual to the proper medication and dosage and the problems involved. I want to talk about schizophrenia and let them know it is not so far removed from them and correct some of the common misconceptions held about people who have schizophrenia.

Let me explain some of the major problems and pressures that schizophrenia has presented to me in getting through pharmacy school.

During my first semester of pharmacy school I was on 2 mg of Haldol [haloperidol] and 2 mg of Cogentin [benztropine, an anticholinergic drug used to reduce such side effects of antipsychotic drugs as speech slurring, rigid neck muscles, and fixed gaze] h.s. [at bedtime] as prescribed for me after hospitalization the summer before entry to school. My condition improved psychologically and I seemed to be in remission until I entered school. I found I could neither read the board nor my notes; everything was blurred no matter where I sat. I called the psychiatrist who had prescribed the drugs and remembered his suggesting that I should take 2 more mg of Cogentin. (It is not clear whether the psychiatrist misunderstood my reason for calling or whether I misunderstood his advice about Cogentin. However, I later learned that Cogentin makes blurred vision worse, not better.) I complied and the next few days I not only had blurry vision, but I could not even see the lines on my notebook paper nor my writing — it was all one blur. In fact, the paper looked colorless. After 2 to 3 days of this, I called the physician back and told him I just could not take this medicine any more, because I could not read or see with it. I could not even tell if I was taking notes on the lines. This side effect, he said, was as he expected; his recommendation now was to drop down to only 2 mg of Cogentin and switch from Haldol to Stelazine [trifluoperazine] 6 mg every day h.s.

This was a compromise solution because, although I could now read and write, my schizophrenia was not so well controlled. I wanted to drop out of school 3 weeks into the semester; I was afraid to go outside and felt as though I did not belong in pharmacy school or would not be able to overcome the stresses to be faced there. Fellow students were remarking to me that I seemed to be more impatient, hyperactive, and depressed. I also had problems with what a friend of mine called "the Stelazine stroll" — akathisia [restlessness]. I continued to go out of the city once a week to see my psychologist, who helped me with aspects of the pressures I could not face alone or only with the drugs.

In an effort to be self-destructive, and perhaps as part of the uncontrolled disease process, I stopped my psychotherapy and medication for 3 months. I was a pharmacy student with probably one of the worst compliance problems possible. I should have known better, but the intellectual knowledge I had gained was applicable to everyone except myself. I was not even at this point entirely aware of why I was so noncompliant except that I was self-destructive. The schizophrenia worsened and I became depressed in addition, due to my inability to cope. After pushing myself to the point of a psychotic break, I finally called my psychologist 3 months later and leveled with him about not being on medication. He worked with me about starting it up again. It was now the second semester of my first year in pharmacy school, a semester in which the schizophrenia remained in control with therapy and medication.

Summer school started and I was doing relatively well until there was a personal crisis to which I responded by going off medication. After all, all these other people around me were making it without meds, why couldn't I? So I went on and deteriorated during the summer and into fall until the choice became 4 to 8 weeks of hospitalization (and dropping out of pharmacy school as a consequence) or taking the medicine. I chose the latter, saying to my psychologist that I didn't like either choice.

I was then in my first semester of my second year. I had just restarted Stelazine at 8 mg h.s., an increased dose, and began having what I thought were seizures. In my classes I experienced an aura and then a wave hit me. I felt overstimulated and could hear a lecture but not process the information and take notes. My hand tremor was so bad during these episodes that I could not write. My psychologist suggested a consultation with the psychiatrist who had supervised my previous hospitalizations and prescribed the medications.

Although the psychiatrist was hesitant to give me the label for what was happening, I insisted, and he said it was "transient psychotic episodes." The problem with this development was that it began after I had already been taking an increased amount of the medication. Where could we go from here? The psychiatrist recommended titration, increasing the dose of Stelazine. However, it didn't work. He then suggesting taking Stelazine along with another antipsychotic drug with more milligram potency (Navane [thiothixene] 5 mg h.s.), but I was still having acute psychotic episodes in my classes. I had taken to sitting in the back of the classroom, although I could not see the board, because I needed to be able to leave the room when this occurred, at the suggestion of the psychiatrist that I not sit there and suffer through it. I had explained away my change in seating to the other students by saying I felt I was going to have a seizure or by joking and saying I had decided I didn't care to see what teachers were writing on the board any more.

When I got up in the morning, I could predict that the episodes would occur and where — I had a prodrome [a premonitory symptom]. There were many frantic long distance calls to my psychologist after these episodes. I had to tell someone who could help me with what was happening to me. I, at this point, felt scared enough that never again would I have a compliance problem. I didn't want to lose all I had worked for in pharmacy school. I noticed the episodes were worse when emotionally volatile material was discussed in classes, such as antipsychotic agents, characteristics of schizophrenia, depression — all problems I had to cope with daily and that remained unresolved for me. . . .

As a consequence of the psychotic episodes and occasionally having to leave the classroom, I missed a lot of notes in my classes. All this work had to be made up. This increased the pressure I was under, which in turn worsened the schizophrenic symptoms and almost forced me into hospitalization. I did not want to drop out of school or receive too many incomplete grades, which would have been the result of 4 to 8 weeks of hospitalization to get properly titrated on the medication and to decrease disease symptoms. However, most of my instructors had rigid rules about missing exams and taking make-up exams. To reduce the pressure, I told the professor with whom I was doing independent study that for medical reasons I would not be able to finish the paper due in that course. I decided to tell him why and he allowed me the Incomplete grade without requiring a medical letter on file, saving me the possible consequences of having this information on written record. And, most importantly, he did not treat me differently as a result of knowing. This reduced my stress and gave me time to make up work and take my final examinations. It also allowed me to work on my independent study paper during vacation and to do a good job on it while I was finally beginning to get a positive response to the medication.

I enjoyed winter break and finished my independent study project without incident, but as second semester approached I began to fear the room we had classes in, all the people and stimulation, and to fear recurrence of these episodes. What scared me most was the fact that this disease could prevent me from doing something I really wanted to do and needed to do, be psychologically healthy — that is, complete pharmacy school — and the knowledge that schizophrenia does this to many people's lives. I could not accept the fact that intellectually I could be capable of something that I may not at times be capable of emotionally.

When classes started, I still felt overstimulated and again had prodromes of psychotic episodes. I could not process information when people were talking; everything just seemed like noise. I was now on 5 mg of Navane b.i.d. [twice daily] and 2 mg of Cogentin h.s. I got up enough courage to sit in front of the class again, but I was very fearful. My psychologist explained that I had begun to associate that classroom with these episodes and that extreme anxiety was causing dissociation reactions in me: I felt I was outside my body; I was watching everything. I wanted an antianxiety agent to get rid of these feelings and that constant impending feeling that a psychotic episode would begin. The psychiatrist prescribed 5 mg of Valium [diazepam] in the morning and at bedtime when necessary. I took it only in the morning when I could not restructure my environment and situation to reduce the anxiety. For the first several weeks I was falling asleep in my first class and had double vision because I could not keep my eyes open. Finally, I became tolerant to the sedative effect.

So this is the answer right now for me: neuroleptic, an antianxiety agent, an anti-Parkinson agent, and intense long-term psychotherapy with my psychologist. And I still look around at my fellow students and say to myself, "They do it without medicine, or doctors, or going to a psychiatric ward," but I needed all these things to cope with the pressure and stress of pharmacy school and life.

What I have been trying to express here is the actual reality of what being "individually titrated to an antipsychotic medication" and having schizophrenia means to someone personally going through it as opposed to how objectively and easily it is expressed in pharmacy classes. My instructors have stated that "antipsychotics alleviate symptoms but do not cure psychoses," but this matter-of-fact statement has very personal meaning for me. It involves internal conflicts and many complicated adjustments — getting to a psychologist outside the city, or if the necessity of hospitalization occurs, getting hospitalized outside the city so fellow students and the pharmacy school will not have access to that information about me. It means never being able to see well because of the side effects of the medication. It also means enormous medical bills and debts.

I recall a teacher, a Pharm. D., telling the class that schizophrenics tend to have low IQs. He was wrong; the research does not support this. They probably do tend, because of the disease, to be more environmentally deprived and have interruptions in their schooling.

I have heard fellow students talking about violent crimes saying, "Oh, you know that person was schizophrenic." No one is teaching these health professionals what the word means, what it does to people, and that schizophrenics are generally less violent than the rest of the general population.

Finally, I heard a teacher in one class talk about long-term chronic illness such as schizophrenia in a way that suggested the teacher knew something about the disease and had looked beyond the myths. Through this class, I began to understand a little better my own noncompliance with the psychotropic drugs; how unacceptable my illness was not only to me, but would have been to others if they had known my diagnosis. I didn't take the medicine at times because I didn't want the disease, its problems, and its stigma. I wanted to be normal. And even now in the 1980s, in a professional pharmacy school, it would probably shock many people to know a schizophrenic was in their class, would be a pharmacist, and could do a good job. And knowledge of it could cause loss of many friends and acquaintances. So even now I must write this article anonymously. But I want people to know I have schizophrenia, that I need medicine and psychotherapy, and at some times I have required hospitalization. But, I also want them to know that I have been on the dean's list, and have friends, and expect to receive my pharmacy degree from a major university.

When you think about schizophrenia next time, try to remember me; there are more people like me out there trying to overcome a poorly understood disease and doing the best they can with what medicine and psychotherapy have to offer them. And some of them are making it.

Reprise

You have now been exposed to a selected sample of oral histories from your fellow human beings who happened to experience schizophrenia personally. Many other stories were omitted because they have not yet been written or because the individuals involved are too ill, bitter, or confused to speak. Certainly, the educational attainments of the story writers here assembled are atypical. Suggested readings for this chapter provide additional and more detailed personal accounts of this tragic disorder.

4

Schizophrenia across Time and Space

Epidemiology and Demography

Providing details for the two domains in the title of this chapter is crucial to furthering our ultimate objectives of discovering the causes of and the cures for schizophrenia. The paths to the answers are very much like those followed by Sherlock Holmes in his successful efforts to bring a guilty culprit to justice despite a bewildering array of valid as well as misleading clues in his path. Careful and precise description must precede measurement and quantification to ensure that the pieces unearthed are those for the schizophrenia puzzle and not for some other puzzle. The particular and the holistic approaches to the description of schizophrenia have been provided in the previous two chapters. So armed, we can now proceed to measure the frequency of the disorder and to localize it in space and time.

When investigating the patterns of appearance of a disease in populations, it quickly becomes obvious that schizophrenia and other health-related disorders are not distributed randomly. Once the true rate of schizophrenia in the general population is established as a benchmark, departures from that mark, either excesses or shortages, command our attention as clues to the causes of and/or risk factors for the condition. Before we can determine the causes of schizophrenia, however, very specific studies are needed to identify the probable environmental and/or genetic factors associated with the observed excesses or shortages in the rate of schizophrenia. Some aspects of the epidemiology (the study of the incidence, distribution, and control of a disease) of schizophrenia have been well studied, although they require reevaluation—or, perhaps, redoing—in the light of the advances in diagnostic accuracy described in Chapter 2.

Descriptive epidemiology has yielded, for instance, rates of schizophrenia by age, sex, social class, education, and nationality; genetic epidemiology has yielded rates as a function of genetic relatedness of persons to a known schizophrenic; and ecological epidemiology has yielded rates as a function of neighborhood (e.g., slums versus suburbs), urban versus rural living conditions, degree of exposure to industrialized versus agrarian lifestyles, and season of birth. The data gathered from these and other studies permit us to focus our attention on "host" factors, such as the genotype or the physical condition of the patient; or, alternatively, on environmental factors (physical and/or psychosocial); or, ideally, on the interaction of both host and environmental factors within an enclosed, interdependent biological-psychosocial system.

Lessons from Epidemiology

In scientific sleuthing, the study of the incidence (the rate of appearance of new cases) and distribution of a disease has sometimes turned up the key to finding a cause. For example, understanding the causes of cholera was set in motion in 1854 by John Snow, who noticed the neighborhood patterning of the disease in a region of London where two wells supplied all the water; removing the pump handle from the well where the cholera victims had obtained their water led to a dramatic reduction in the incidence of new cases. However, the evidence for an infectious "germ" from sewage-contaminated water that caused the spread of cholera was only circumstantial. Scientists had to await the microscopic analysis of bacterial cultures by Robert Koch in 1883 to actually see the bacillus (he had discovered the cause of tuberculosis—another bacterium—a year earlier).

In the 1930s, many psychiatric hospital beds in the southern United States were occupied by people, more often black than white and much more often poor than rich, with a strange psychosis; it was said to affect

20 percent of black admissions to one mental hospital in North Carolina. Although the disease resembled an organic psychosis, it could imitate the picture of catatonic or paranoid schizophrenia. Epidemiologic detective work by Joseph Goldberger had identified the disease by 1915 as nutritionally caused *pellagra*; that it was caused specifically by a deficiency of the B vitamin niacin in people whose diets were heavily dependent on unprocessed maize (corn) in their diets was only discovered in 1938 by Tom Spies. The use of niacin (nicotinic acid) therapy, together with massive improvements in the diets of the poor, including the introduction of vitamin-enriched bread in the 1940s, has virtually eliminated this disease in developed countries. Once the puzzle pieces were fitted together, false ideas about causality related to personal characteristics of the "hosts," such as racial or class inferiority, were definitively refuted.

A more recent example of the power of experimental epidemiology to influence the search for causes comes from the study of AIDS. Although originally perceived as a disease of homosexual males, further data about drug addicts, heterosexuals, babies of affected mothers, and recipients of blood transfusions greatly expanded the number of viable etiological hypotheses that scientists began to pursue, with rapid results.

We are in dire need of studies that yield more specific rates of schizophrenia among social classes, among those who have undergone various psychological stressors or specific physical insults to the brain, and among ethnic groups protected from, and exposed to, specific putative causes. We need, as well, more specific environmental studies, difficult as these are to evaluate and control, before we can judge the too numerous claims about the role of diet, viruses, parenting practices, and "civilization" and can test their roles, if any, in the development of schizophrenia.

Knowledge about the causes, treatments, and outcomes of schizophrenia advances by the defeat and the victory of various hypotheses when they are confronted by well-documented facts. Thomas Huxley once commented that "tragedy in science is the slaying of a beautiful hypothesis by an ugly fact." We too must insist on such a confrontational use of data to avoid the comfortable complacency that may come from embracing a so-called biopsychosocial model for the causes of schizophrenia, diplomatic as that may be. Of course, such a model is better than are narrow-minded genetic, psychodynamic, or sociological ones. Our goal is to distinguish between a condition that is, say, 5 percent genetic and 95 percent environmental (such as tuberculosis) with regard to the contributors to its development, and one that is quite the opposite (such as diabetes or coronary heart disease), so that we can use our resources rationally. Agreeing that genes and environments are both important in causing schizophrenia may be analogous to innocent acceptance of a recipe for mastodon and rabbit stew that calls for 50 percent of each and then discovering, on further inquiry, that the chef meant one of each species.

Is Schizophrenia a Recent Disease?

Would that we had careful, documented historical studies of schizophrenia! They could tell us a lot. For instance, if there were marked changes in the incidence of new cases of schizophrenia specifically, not just of "lunacy" in general, in western Europe between 1800 and 1900, we'd suspect that the social and personal stresses accompanying industrialization and urbanization played a causal role or that somehow there had been a change in the frequency of relevant genes. If there were changes over a shorter period, we would look for the aftereffects of a viral epidemic, a newly mutated virus, a marked change in child-rearing practices or in mental illness "head-counting" practices, or some other abrupt change in society such as the addition of vitamins to our milk or fluoride to our water.

Fascinating as these possibilities may seem, they are all hypothetical. From the research reviewed in Chapter 1 on the historical roots of schizophrenia and in Chapter 2 on the effects of fashion and varying criteria on who is diagnosed as a schizophrenic, it is obvious why older epidemiological surveys and anecdotal evidence yield ambiguous findings. Dr. E. Fuller Torrey of Saint Elizabeths Hospital, Washington, D.C., has collected much of the relevant literature in his *Schizophrenia and Civilization* (1980), and Dr. Edward Hare, retired psychiatrist and scholar from the Royal Bethlem Hospital, has scrutinized the historical record as well (1988). But can the data available be taken at face value? Torrey finds little to support psychosocial, psychodynamic, or sociocultural theories of causation. *Something* biological and/or genetic is compatible with the reported distribution of schizophrenia over time and place. The changes in frequency noted over the past 200 years are ambiguous — some observers believe that they reflect real changes in incidence, others believe that they are simply artifacts of changing diagnostic criteria, availability of institutions, and reporting methods. Inability to document changes in disease patterns, however, doesn't necessarily mean there are none. It may mean that we are the victims of poor record-keepers and the former state of the art for studying aberrant behaviors.

Official census data for England and Wales for 1859 show that there were 36,480 "lunatics, idiots, and persons of unsound mind" among a total population of almost 20 million; by 1899, the population had grown to 32 million, while the "deranged" population had increased at a much higher rate, to 103,247. Without some substitute Rosetta Stone for translating nineteenth-century "psychiatrese," the data are difficult to use in support of theories about acute changes in the incidence of schizophrenia and, hence, its causes. Even as recently as 1914, the census for England and Wales reports, as a yearly average for the period 1909–1913, a total of 15,150 "functional psychoses" among a grand total of 21,832 "lunatics" admitted to all lunatic asylums. The category of functional psychoses was made up of the following: stupor, primary and secondary dementia, mania, melancholia, alternating insanity, and delusional insanity — tantalizing but frustrating information for the "paleontology" of schizophrenia.

From Raw Data toward Strategic Information

In making use of population-based data, we need to be aware of how easily statistics, whether historical or contemporary, can mislead the search for clues to causes. In the decade between 1965 and 1975, the total resident population of state and county mental hospitals in the United States was reduced by 60 percent. In the shorter interval from 1969 to 1975, the number of resident schizophrenics in these hospitals dropped from 184,000 to 93,200. This decline did not mean that there was an abrupt change in the frequency of schizophrenia; it simply meant that changing American health policy decisions had significantly altered how and where schizophrenic patients were treated. The mental health delivery system changed drastically beginning in 1967, with the well-intended but misguided policy of accelerated deinstitutionalization. The schizophrenics who were formerly hospitalized are now seen and counted (or missed) in different settings. With help from improvements in antipsychotic drugs, many schizophrenics have returned to the community. Some are in supervised public settings, and we still are able to count those, but some are in jails and prisons, others are lost among the hordes of homeless, and a few are in private treatment. It is quite difficult to get accurate estimates of the numbers of schizophrenics when they are not in the established mental health "territory." The impact upon society of the reduction of the total state and county hospital patient population from

550,000 in 1955 (the peak) to 111,000 in 1986 — an 80 percent decrease — was revolutionary and still is seldom acknowledged. The number of homeless people in the United States as of 1988 is estimated (Torrey, 1988) to be 450,000; even after excluding those primarily affected with alcohol and drug problems, 150,000 appear to be seriously mentally ill, and exactly how many of them are schizophrenics is not known. Yet another tragedy is the fact that the current number of persons incarcerated for legal offenses in jails and prisons is about 700,000; from 5 percent to 20 percent are estimated to suffer from serious mental illness, but how many of them qualify for a diagnosis of schizophrenia is not known.

As a first step to determining the magnitude of the problem of schizophrenia, Table 5 gives official 1989 government statistics for total admissions to and residents of U.S. mental health facilities in 1986, the most recent data available. Schizophrenics are singled out for comparison with the total, which includes patients afflicted with depression, mania, organic brain diseases, and alcohol and drug disorders. These statistics, however, underestimate the true number of schizophrenics in the population because of some important exclusions beyond those just mentioned. They do not include persons who were treated previously but were not then under care, and they exclude not-yet-diagnosed and untreated cases in the general population.

Even if we had those numbers, figures such as those shown in Table 5 can't be used by themselves to estimate the rates at which schizophrenia appears in a population, although they are essential for planning delivery of mental health services. We would need to know how the 883,290 schizophrenics known to be under care correspond to the total U.S. population. Then we can estimate the prevalence rate of schizophrenia in the general U.S. population. We can also convert this kind of information into estimates of age-corrected morbid risks (MR) — the likelihood that someone will suffer an episode of schizophrenia between birth and death. For schizophrenia, these ways of counting have been developed into some standardized indices.

It's always important to know how information for a study was gathered and, of course, who was included in both numerator and denominator. Generally, numerator information comes from two methods: door-to-door field surveys or records of contact with relevant institutions (hospital admissions, for instance). With schizophrenia, the results of these two information-gathering systems are remarkably similar. This is not necessarily true of other disorders, even other psychiatric disorders, as we'll see from some classic studies.

In addition to lifetime MR, there are two other counts that are important in the study of schizophrenia: *incidence* and *prevalence*. The

TABLE 5

All known psychiatric patients under care* and those with a diagnosis of schizophrenia in the United States in 1986, by organization

	Total Number under Care	(%) Schizophrenic	Number of Schizophrenics
Totals	5,976,326	15%	883,290
INPATIENT PROGRAMS	1,884,463	21%	396,989
State and county mental hospitals	456,142	36%	163,369
Private hospitals	235,264	8%	18,961
VA medical centers	199,186	25%	49,696
General hospitals (separate psychology)	893,091	15%	136,977
Multiservice organizations	100,780	28%	27,986
OUTPATIENT PROGRAMS	3,784,457	10%	394,089
State and county mental hospitals	100,884	26%	26,718
Private hospitals	132,925	6%	8,413
VA medical centers	134,242	20%	27,030
General hospitals (separate psychology)	502,537	10%	52,781
Multiservice organizations	2,206,746	10%	229,671
Residential for children	34,501	6%	2,211
Freestanding clinics	672,622	7%	47,265
Partial care programs	286,784	32%	92,213

*Combines all present on a single day with those admitted over the entire year of 1986.
Source: Adapted from Rosenstein et al. (1989).

incidence of schizophrenia is the number of new cases that occur within a period, usually one year, expressed as a rate per 1,000 or 10,000 of the general population. Incidence rates may be suspect because they are sensitive to variable diagnostic standards and to the availability of official service delivery points for detection of mental disorders. They are most helpful, however, when they can be made age-specific. The Communicable Disease Center in Atlanta, Georgia, uses information on age-specific incidences to track changes in the frequency of AIDS and tuberculosis as epidemiologic clues to contributors and causes (and the need for services), just as we do for schizophrenia. The summation of age-specific incidences over the life span also provides a shortcut equivalent to the lifetime MR, without requiring researchers to wait 55 years to gather the information directly from the longitudinal observation of a birth cohort.

The *prevalence* of schizophrenia is based on the total number of patients alive at any specific time; this number would include those who have been ill in the past, but who are not obviously psychotic and under "official" care at the time of study. Information about such remitted cases or not-yet-diagnosed cases depends on both the accuracy of psychiatric case registers (longitudinal, regionally based record collection) and the validity of door-to-door psychiatric surveys. Prevalence is expressed as a rate per 1,000 either for the total population or for only the people over the age at which the risk for developing the disorder begins, usually fixed at age 15 for schizophrenia. It is important to know which of these two denominators is used to calculate prevalence. In the United States and Western Europe at present, 70 percent of the population is over age 15, so general population prevalences are much lower than adulthood-specific prevalences. (The numerator stays constant, but the denominator increases dramatically, so the percentage is much smaller.) One needs to look very closely at any statistics on schizophrenia to know what they represent. It would be incorrect, for example, to take at face value the prevalence of schizophrenia for the total population of Mexico where only 57 percent of the population is over age 15, or any other developing nation with high birth rates, and to compare it with the same figures for Denmark or the United States. Prevalence based on the total population of Mexico would be much lower than the same figure for the United States, since it contains a much smaller proportion of adults who could be affected with schizophrenia; taken at face value, such a lower rate would be a false clue and would mislead one to look for something protecting against schizophrenia in a country like Mexico, or something exacerbating it in the United States.

For genetic research and counseling, we need still more information. If we are to know how risk varies as a function of differing degrees of family relatedness, we first need to know the lifetime MR in the general population. For this, we rely upon already developed indices, the most commonly used of which comes from Wilhelm Weinberg (1862–1937), a German obstetrician, pioneer of medical genetics, and adviser to Ernst

Rüdin, founder of the Munich school of psychiatric genetics. In Weinberg's MR index, the denominator is reduced to allow for the ages of the subjects much as are actuarial tables for calculating insurance premiums. The age determines how much of the total lifetime risk for developing schizophrenia has been "used" and how much is left (see the next section). The great advantage of Weinberg's method for geneticists is that it allows simultaneous comparisons across generations — for example, risk in grandparents compared to risk in grandchildren — even though they differ greatly in their ages at one particular time.

To comprehend the information on the risk of developing schizophrenia in the relatives of schizophrenics, it is important to understand in a bit more detail both the age of onset and the role that longevity plays in calculations of prevalence and risk.

The Age of Onset of Schizophrenia

Systematic observations since the time of Kraepelin reveal that schizophrenia has a variable age of onset: Almost no cases appear before the age of puberty; the number of new cases sharply increases between the ages of 15 and 35 and then slowly drops off; very few new cases appear after age 55. The variable age of onset for schizophrenia, together with the average length of life of a population (a topic we usually ignore in the 1990s), has important and direct implications for the number of individuals we can expect to observe affected by schizophrenia at any point in a life span or historical time. If people die before the possible age of onset of the disorder, subsequent historical changes in longevity will lead to an increase in the observed number of cases: Alzheimer's disease was not observed among cave men and women because few, if any, lived long enough to show the symptoms.

Half of all male schizophrenics experience the onset of their illness by age 28 and for females, the age is 33, according to reliable national Norwegian statistics (Saugstad, 1989). Representative average ages of onset for DSM-III schizophrenics were lower: 21 years for males and 27 for females in the careful studies conducted by Armand Loranger at Cornell Medical Center in New York.

The sex difference in age of onset shows up around the world and represents an enigma. The total lifetime morbid risk does not differ significantly by sex. Can the sex difference at onset be attributed to differences in the age at which the brain matures, to the greater vulnerability of males burdened with testosterone, to the protective effects of estrogen in females, to greater social demands on males to "grow up" before they are as socially and physically mature as their female age mates, or what? We don't now, but efforts to find the answers to such questions will certainly add to the knowledge needed to solve the puzzle of schizophrenia.

We can now tabulate the accumulated lifetime MR of being admitted to a psychiatric hospital with a diagnosis of schizophrenia by age x, for males and for females. Table 6 draws upon first-admission data for England and Wales for the years 1952–1960, after the advent of national health insurance with very accurate record-keeping. It also shows the proportion of the total risk used up by age x, assuming that the end of the risk period for developing the illness is age 55. Thus, the age-specific risk for males reaching the age of 25 is only one-third of 1 percent, not the 1.06 percent risk for the entire population of males who manage to reach age 55. By age 40, males have used up 81 percent of their lifetime risk,

TABLE 6

Age-distributed risk of first hospitalization for schizophrenia by sex in the British population

By Age	Accumulated Risk (%)		% of Total Risk Used Up	
	Male	Female	Male	Female
15	0.01	0.01	1	1
20	0.12	0.10	11	11
25	0.33	0.25	31	25
30	0.54	0.42	51	44
35	0.75	0.59	70	60
40	0.86	0.71	81	72
45	0.97	0.82	91	83
50	1.02	0.91	96	92
55	1.06	0.99	100	100

Sources: National data for England and Wales, 1952–1960, from Registrar General, 1960. Adapted from Slater and Cowie (1971).

whereas females have used up only 72 percent of theirs and still have 28 percent of it left.

The Demographics of Longevity

Demography is the statistical study of changes in population number and distribution; it examines such variables as fertility, death rate, marriages, inbreeding, and migration, as well as age and sex specificity for such variables. No science of psychiatric demography exists as such, but contemporary psychiatric epidemiology approximates it. Tombstone inscriptions, tax records, and even telephone books can provide the raw data for demography.

Look at Table 6 and contemplate the effect on the number of schizophrenics to be observed in a population if everyone were to die by age 25. The number of schizophrenics would drop by 70 percent! Figure 4 plots the average life expectancy at birth from ancient times to the present; the estimates come from historical demographers, who often use crude data, such as the estimated ages of small samples of skulls from graveyards. The tremendous leaps in longevity and the consequent increases in population sizes have originated from marked decreases in the death rates from serious epidemics (plagues, smallpox, cholera, typhus), better nutrition, and improved sanitation, all beginning in about the last quarter of the eighteenth century. (William H. McNeil presents a fascinating panorama of the impact of disease on world history in his *Plagues and Peoples*.) Of course, many people lived into their 50s and 60s even in ancient Greece and at the time of Christ in Rome, but they were the exceptions. Figure 4 reveals that the average length of life was only 18 to 22 years at those times, rising rapidly to the present-day values of 71 years for males and 78 for females in the United States.

We can visualize the actual number of persons in a cohort of age mates alive at each age who survive up to each succeeding age under the living conditions prevailing at the time the cohort is born. Figure 5 presents such survival curves for cohorts of 100,000 males: one for rural England for the year 1700; one for the United States on the eve of World War II; and, showing some improvements, for the United States in 1967. The contrast between the England of 1700 and the United States of 1967 is obvious and striking: By age 30, 52 percent of the English cohort had died (30 percent by age 5), compared with only 6 percent of contemporary males. The fact that longevity was comparatively short in the past is certainly a major reason that few cases of schizophrenia-like psychoses were observed before the early 1800s.

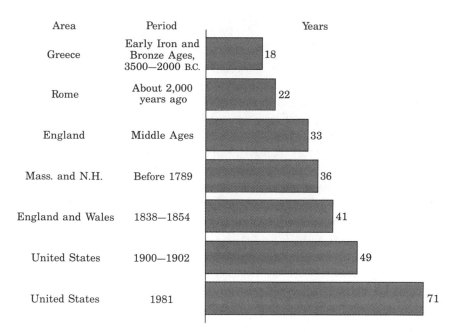

FIGURE 4. Average life expectancy from ancient times until the present. Adapted from Dublin et al. (1949).

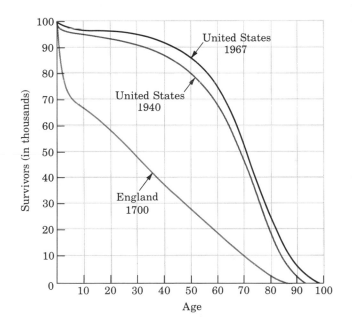

FIGURE 5. Male survivorship to successive ages from a starting birth cohort of 100,000: United States whites, 1939–1941 and 1967, and England, 1700. (Data from Cox, 1970; Dublin et al.; 1949, and U.S. Bureau of the Census, 1988.)

Attempts to Determine the True Number of Schizophrenics

In 1952, the late Norwegian social and genetic psychiatrist Ørnulv Ødegaard, the founder of Norway's national psychiatric case register, questioned incidence and prevalence figures for mental illness that are based only on institutional contacts: Are they so inaccurate as to be almost useless? How many cases of mental disorder would be found by a complete census of the entire population that would not be found by a mental-hospital-patient census? To answer these questions, he devised a formula for estimating a complete field-census count. He knew from previous research that social class, distance of residence from a psychiatric hospital, age, urbanization, and diagnosis all jointly determined hospital use. His information, however, also indicated that despite these factors, schizophrenics were more likely to be hospitalized than were sufferers from other kinds of mental illness, such as recurrent depressives who could be managed at home.

Ødegaard reasoned that even if all schizophrenics were hospitalized, such a prevalence survey would miss those who were suffering from initial symptoms but had not yet been hospitalized, and he estimated that number from his previous experience. In addition, his model provided for inclusion of three other categories often missed in hospital prevalence surveys. He included figures on schizophrenics who would die after onset but before hospitalization, those who had been released, and those who were hospitalized but died before the survey. Thus, he modeled the hypothetical ultimate census.

His data and model yield this picture: A single day's count (a point prevalence) of citizens based only on hospital admissions data would turn up only 56 percent of Norway's total psychotic population. But when only patients with schizophrenia are considered, the in-hospital cases constitute 72 percent of the projected true number. If you add the estimated 6 percent who were formerly hospitalized, and thus officially recorded, but who had been released before the survey, the in-hospital census finds fully 78 percent of the projected number of schizophrenics. Thus, a hospital-based count would have missed only 22 percent of the true prevalence of schizophrenics. In addition, with few exceptions, almost all of these would be hospitalized and counted eventually if they survived.

For manic-depressive disorders, the story is quite different. Ødegaard's statistics show that only 40 percent of these people would turn up on a hospital point-prevalence census, so more cases would be missed than found. Adding former manic-depressive patients would bring this figure to only 68 percent, meaning that nearly one-third of the manic-depressive living in the general population had not been hospitalized; although, like schizophrenics, some of them would eventually be hospitalized if they survived.

As a result of his findings, Ødegaard concluded that, for major mental illness, a field survey was superior, but not vastly superior, to a hospital count, because most persons suffering psychoses eventually were hospitalized. For schizophrenia, he thought that almost all those afflicted would be hospitalized at some time. The conclusions—though heartening for researchers, because relying on hospital statistics is far easier and cheaper than is conducting field surveys—only underline how serious and how disruptive schizophrenia is and why we must pursue the quest for causes and rational prevention. They tell us that schizophrenia is too serious to be ignored, no matter what the circumstances. Given the figures cited above for the numbers of homeless and for those incarcerated in penal institutions, together with the fact that this nation lacks the socialized medicine safety net of the Scandinavian countries, it would be unsafe to generalize from the adequacy of the Norwegian statistics to the adequacy of hospital-based head counts conducted in other kinds of national systems of health-care delivery.

From Mental Hospital Statistics to Community Estimates

Let's look at two classic studies in psychiatric epidemiology that were based on field-census methods rather than relying only on cases of schizophrenia that appeared in a hospital before they could be counted. Both are Scandinavian, and both undertook a broad mental health survey in which schizophrenia data were reported separately.

In 1947, Erik Essen-Möller and three psychiatrist colleagues conducted a careful door-to-door census of the 2,550 men, women, and children in rural southern Sweden—all the *living* inhabitants of two parishes. Virtually everyone was interviewed, and information gathered from them was complemented by data from official sources—schools, tax authorities, alcoholism registers, mental and penal institutions, and so on—and from family physicians. The data are a gold mine of information.

The team found 17 definite schizophrenics and 4 probable schizophrenics; knowing Essen-Möller's diagnostic standards to be conservative from his handling of the Maudsley twin sample (see Chapter 2), we are inclined to count the probables as real cases. Of these 21, 6 had never been hospitalized and 10 were actively psychotic at the time of the interviews, but only 5 of these were hospitalized (2 more should have been and 3 were in other protected settings). Ten schizophrenics were gainfully employed. Thus, a current hospital-based point-prevalence estimate of the frequency of schizophrenia would have missed 16 of the 21 true cases in the community.

Researchers would like to take comfort in the argument that a point prevalence based on hospital admissions simply misses many nonhospitalized but mildly schizophrenic cases or "spectrum" (schizophrenia-like personality disorders) relatives of schizophrenics. They know that some of these cases would be identified in a follow-up study, but in general they hope that correcting for age by using the procedures described above can compensate for errors that creep into the numerator.

To see the difference in terms and methods of counting, let's look at Essen-Möller's statistics. If 17 definite schizophrenics are found in a general population sample of 2,550, then the lifetime prevalence (17/2,550) is 0.67 percent. If we add in the 4 probables, the prevalence is 0.82 percent. If we consider only the population over age 15, the denominator is reduced and the 21 schizophrenics yield a lifetime prevalence of 1.1 percent. When the denominator for the Weinberg abridged method is age-corrected with a risk period defined as age 15 to 40 (preferred by Essen-Möller for his generations), the lifetime morbid risk is 1.12 percent. If the 4 probable schizophrenics are included, the lifetime morbid risk rises to 1.39 percent. This last figure is the most accurate reference value for comparing risks observed in the relatives of schizophrenics in that part of Sweden to the risks observed for a member of the general population at that time. Only when the risk in relatives significantly exceeds the general population risk are we justified in concluding that schizophrenia is indeed "familial."

In principle, the most accurate information about any disease's distribution would come from taking a newborn infant population of considerable size and following and accounting for the individuals in that population for the next 55 years — obviously a formidable task. Kurt Fremming, a Danish psychiatrist, conducted one of only three mental health studies that use this design. Fremming's *biographical* or *cohort* method differs from Essen-Möller's *census* method chiefly in that Fremming counted the schizophrenics who died over the course of the follow-up period as well as those still living at the end of it: None of his subjects were older than 59. He began with 5,529 persons born between 1883 and 1887 on the island of Bornholm in the Baltic Sea between Denmark and Sweden and then traced them up through 1940. He discounted those who did not survive to age 11, leaving 2,120 males and 2,010 females to make up his data set. The lifetime morbid risk for schizophrenia that emerged from this biographical method was close to 1.0 percent. The risk was derived from a numerator of 38 schizophrenics who were found in the population followed from age 11; without the 13 of these who had died during the course of the survey, lifetime prevalence would have only been 0.62 percent for this Danish population and, at first glance, much lower than the values for lifetime morbid risk of either 1.12 percent or 1.39 percent for the Swedish sample above. It is obvious that allowing for schizophrenics who have died and dropped out of sight leads to a more accurate estimate of risk and to a more accurate benchmark value. By comparing the general population risk of 1.0 percent from the biographical method with the risk of 1.12 percent or 1.39 percent from the census method, we can conclude that they converge upon a more or less equivalent value; henceforth, we can depend upon the more practicable census method.

The value most often accepted as representing the lifetime morbid risk of schizophrenia in the general population is 1.0 percent. That is, 1 out of every 100 people born today and surviving to at least age 55 will develop into a diagnosable case of schizophrenia. The earliest census studies (1928) coming out of Germany had established the risk at 0.85 percent based on a very small sample, and that figure is used in the older literature. The currently accepted value is not, after all, so far from the "lucky" guess.

It is important to clarify what the lifetime risk of developing schizophrenia means and what it does not mean. It means that, for the purposes of research into the causes of schizophrenia, a "normal" rate of schizophrenia is considered to be 1 percent, and any higher occurrence in any studied group must be due to some increase in risk factors — perhaps an inherited component, an extremely stressful environment, injuries, or illnesses. If, for instance, research indicated that the population of rural western Ireland had a lifetime risk of 4 percent, we would want to ask what contributing factor was more in evidence there than in London or Honolulu. Or if there were a significant difference in risk among ethnic groups living in the same urban area, we would attempt to isolate some ethnic-specific contributing cause.

A 1 percent lifetime risk for schizophrenia does not mean that every individual has that risk. To arrive at that average, we counted individuals from families who in five generations have never had a case of schizophrenia occur, as well as individuals born of two schizophrenic parents. The individual who has had no schizophrenic relative for generations has an even lower than 1 percent risk, whereas the individual who has two schizophrenic parents has a lifetime risk for developing schizophrenia of between 36 percent and 55 percent, as we will see in Chapter 5.

The Distribution of Schizophrenia by Social Class

Of the many ways of dividing a population to find differential distributions of the risk for developing schizophrenia, and thus to provide clues to the likely causes or contributors, we'll deal here with only one sociocultural division: social class distribution as inferred from occupation, because it has one of the strongest associations with risk. If we can discover high rates of illness in one particular social class, perhaps we can see how else that class differs from others and isolate a cause (remember John Snow and the pump handle connection to cholera), Many studies focus on ethnicity, occupation, marital status, religion, and emigration. Schizophrenia among the western Irish, for example, is an especially fascinating topic because the rates are alleged to be four times higher than expected; but, by and large, this observation has so far yielded no significant clues as to cause and needs to be repeated with modern diagnostic standards.

Ever since a study by the sociologists Robert Faris and H. Warren Dunham in 1939 showed that the first-admission rates for schizophrenia were 102 per 100,000 among those living in the slums of central Chicago and diminished in a gradient to less than 25 per 100,000 in affluent neighborhoods on the periphery of the city, researchers have used the information to support opposite conclusions about the nature and direction of the possible causal factors involved. William Eaton, a contemporary sociologist, reports that similar variations have been demonstrated in 16 subsequent studies in other cities and countries and that social or occupational class, rather than local ecological conditions, has been pinpointed as the determining variable for these gradients. Social class is less powerful in rural areas as a predictor of increased rates of schizophrenia.

Eaton, using the Maryland psychiatric case register, found that rates of first hospitalization for schizophrenia were twice as high in central city areas than in other urban or rural areas; by using occupation rather than social class as a predictor, he found that male blue-collar workers (1965–1966) had rates of first hospitalization for schizophrenia that were five times higher than those observed for professional and technical workers. It is important to note that such a relationship between social class and manic-depressive psychoses (major affective disorders) has not been found; thus, generic studies of "psychoses" will obscure the relationship between social class and schizophrenia.

Two principal explanations for the higher rates of schizophrenia among the "underclass" have been proposed. The *breeder* hypothesis, or stress-as-cause, suggests that the cumulative stress induced by poverty, social disorganization in the form of crime and broken homes, and child abuse and neglect in the lower classes is responsible for breeding more schizophrenia. The *drift* hypothesis suggests that a downward social class migration of mentally impaired (that is, mentally inefficient, not mentally retarded) workers on their way to becoming overt schizophrenics overloads the lower class with schizophrenics. These about-to-become schizophrenics are, in effect, immigrants who carry their predisposition with them and "ruin" the reputation of their new neighborhood; namely, that of the inner-city, blue-collar working class. A complementary hypothesis, supported by data, is that social selection traps some preschizophrenics in the lower class from which they originated, whereas their normal siblings and peers are more likely to improve their status as a result of secular improvements in educational and economic opportunities (compared to their parents' generation).

E. M. Goldberg and S. L. Morrison, working with the Social Medicine Research Unit in London, conducted an ideal study to test these competing explanations. They explored the social mobility of schizophrenic males and their fathers, uncles, brothers, and grandfathers. They discovered that although the fathers were distributed among the social classes very much like the normal reference population census as a whole, schizophrenic sons were markedly overrepresented in social class 5 (the lowest class). They concluded that social drift downward had occurred, because a considerable proportion of the schizophrenics in classes 4 and 5 had not been born or reared in lower class homes. The discrepancies between their class of rearing and their social class at the time they were admitted to a psychiatric hospital can be seen in Table 7: More than half the schizophrenics in a lower social class at the time of admission had been reared in higher social classes.

These researchers then explored the processes by which the downward drift had occurred. Although 29 percent of the fathers were in social classes 1 and 2, only 4 percent of the schizophrenic sons were, immediately prior to hospital admission; and although 23 percent of the fathers were in classes 4 and 5, 48 percent of the sons were. The researchers examined the occupational statuses of other male family members to confirm that these too resembled census expectations; only the schizophrenics were "odd men out." The school attainments of the schizophrenics were not noticeably different from those of their brothers; it was *later* in adolescence, when the disease process seemed to have "turned on," that the downward drift commenced. A look at individual British schizophrenics with a grammar school education (college bound, elite education) and with fathers in the upper class shows a wasteland of corroded talent: One passed four "A"-level exams (evidence of high

TABLE 7

Downward social class drift of schizophrenic patients in relation to
that of their fathers, brothers, and census norms as a test of
drift versus breeder hypotheses*

Social Class	Patients (%)	Their Fathers (%)	Their Brothers (%)	Census Norms (%)
Higher (classes 1 and 2)	4	29	21	16
Middle (class 3)	48	48	56	58
Lower (classes 4 and 5)	48	23	23	27

*Social class of patients at time of their admission. Male first admissions age 25–34
years, England and Wales, 1956.
Source: After Goldberg and Morrison (1963).

scholastic abilities) and is now a lab technician; one showed distinction
in four subjects and is now a truck driver's helper; one passed eight "O"
levels (evidence of excellent attainments) and is now a porter for the
railway. Such sad and dreary tales about the destruction of talent can be
reproduced in other countries as well.

The London findings have been repeated in broad outline both in
Detroit and in Rochester, New York. Support for the downward drift
hypothesis, however, does not disprove a role for stressors, although it
may well mean that stress triggers the development of schizophrenia
rather than causes the disease itself. After all, the vast majority of those
born into the lower class do not become schizophrenics. Conversely, we
assume that some persons predisposed to schizophrenia and born poor
might have escaped the illness had they been able to avoid poverty's
insults to development and psychological integration; for example, low
birth weight is associated with poverty and with central nervous system
damage. If not underweight when born, a person might otherwise predisposed to developing schizophrenia might well avoid such an outcome.
Almost all research confirms that the stressors that may trigger
schizophrenia — interpersonal stress, brain trauma and disease, drug intoxication, and the like — are not unique to poverty. At present, the
alleged stressors all lack a proven one-to-one relationship to schizophre-

nia and appear to be too widespread in the population to explain the causes of schizophrenia.

Schizophrenia around the World

The universality of schizophrenia as we have described it, across time (the past 200 years) and space (from the pastoral villages of Botswana in Africa to industrialized societies), has been challenged repeatedly, often with little high-quality empirical data. In 1988, the World Health Organization reported the completion of yet another major study, conducted by Assen Jablensky, Norman Sartorius, and their worldwide collaborators, that should set many of the speculations to rest. Schizophrenia *is* universal, and the variation across cultures as different as rural Chandigarh, India, and Nottingham, England, is demonstrably modest. Careful diagnoses based on structured interviews with 1,379 first-contact psychotic patients from 13 geographically defined catchment areas in 10 countries provided the information for calculating the incidence, prevalence, and lifetime morbid risk for both broad and narrow definitions of schizophrenia. Clinical diagnoses that did not meet the criteria for one of the objective diagnostic classes accounted for only 14 percent of the total sample. Only 56 percent of the total sample fulfilled the narrowest criteria of all — a computer-derived nuclear syndrome or a Schneiderian first-rank-symptom schizophrenia (see Chapter 2). The graph in Figure 6 shows the age-corrected lifetime morbid risks for both broad and narrow definitions using all the starting cases — 1,379 — for eight cultures. An identical picture emerges if one were to plot incidence rates per 10,000 instead of MR. The risk period used — age 15 to 54 — makes the most sense, given the dramatic improvements in longevity and health mentioned above.

For the broad definition of schizophrenia, variation is indeed noticeable as we move from rural India, with an incidence of 4.2 cases per 10,000 at the high end, to urban Denmark, with 1.5 per 10,000 at the low end. As shown in the figure, the variation in MR ranges from 1.74 percent for rural India to 0.56 percent for Denmark, bracketing the expected value in the literature of 1 percent.

When we shift our focus to the narrow definition of schizophrenia, the variation diminishes, as does the prevalence, permitting a generalization about both the consistency and the universality of schizophrenia. The industrial city of Nottingham, England, had the highest incidence, 1.4 per 10,000 per annum; the lowest was observed in urban Denmark, 0.7.

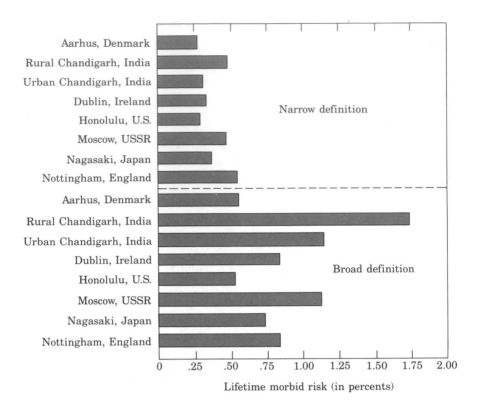

FIGURE 6.   Age-corrected morbid risks for narrow and for broad definitions of schizophrenia from worldwide sites in the World Health Organization multinational study. (Data from Sartorius et al., 1986; Jablensky, 1988.)

The absolute difference is negligible — less than one schizophrenic person in 10,000 inhabitants. The corresponding MRs were 0.55 percent and 0.27 percent, a narrow range. The values reported here are quite compatible with the results from European studies on the epidemiology of schizophrenia after World War II. After reviewing 40 years of experience with the Norwegian national case register, Ørnulv Ødegaard (1971) commented, "The astonishing stability of first admission rates over half a century indicates that in the countries investigated the true incidence of schizophrenia remained essentially unchanged" (p. 56).

It seems safe to conclude that the incidence of schizophrenia in most human populations around the world today is rather similar; even if a rate two or three times as high as the values here is found, it represents a small real difference — 8 or 12 cases instead of 4 cases per 10,000 adults. Still, unexplained pockets of very low rates, such as those reported in the southwest Pacific (Papua, New Guinea) and very high rates, such as those reported in western Ireland and Croatia, Yugoslavia, if confirmed, demand explanations.

Reprise

This chapter has dealt with the many ways of coming to grips with essential questions about the number of schizophrenics in the population, historically and currently, and their location in time and space. Knowledge about the epidemiology and the demography of schizophrenia is necessary to generate and to test hypotheses about the causes of schizophrenia that compete with one another in the scientific arena — not all hypotheses have equal merit, and too much polite diplomacy will erode progress toward the scientific resolution of the mysterious puzzle of schizophrenia.

Of the various indices introduced in this chapter to convey the notion of frequency, morbid risk (the chance that someone born today will develop schizophrenia before reaching age 55) is the most useful for detecting the presence of increased risks in a population or in a sample of relatives of schizophrenics that implicate causes, be they social, ecological, psychological, biological, or genetic. The value of 1 percent as a benchmark value for morbid risk in the population as a whole is worth remembering as we begin, in the next chapter, to investigate the evidence for a major genetic contribution to the causes of schizophrenia.

5

Is Schizophrenia Inherited Genetically?

It should be obvious by now that no such simplistic question as that posed in this chapter's title can do justice to the accumulation of knowledge from research into the complex, multi-factorial causes of and contributors to the phenomenon of schizophrenia. Passionate partisan debate has raged for decades on both extremes of the Nature to Nurture or heredity to environment continuum of explanatory causes; one side asserting that heredity and biology are irrelevant and the other, that environment and psychosocial factors are irrelevant. By the time Kraepelin wrote the final (eighth) edition of his textbook in 1913, he had concluded that the causes of dementia praecox were wrapped "in impenetrable darkness." Such a balanced perspective did not protect the field of schizophrenia research from subsequent extremist pronouncements about the causes of schizophrenia, but three-quarters of a century worth of data gathering have made the darkness much less impenetrable.

The heredity-environment controversy over the causes of schizophrenia is a remnant from a broader, sociopolitical and ideological battle known as the Nature versus Nurture controversy, which dates back to

1859, with Charles Darwin's heretical challenges to the Establishment about the origins of the human species (see Arthur Caplan's *The Sociobiology Debate*). Sir Francis Galton, Darwin's cousin, extended — some would say overextended — the ideas of the evolution of physical characteristics to the inheritance of behavioral traits in a series of papers written between 1865 and 1875. Among the corpses from previous skirmishes may be found the scientific and religious opposition to Darwin's theory of evolution, as well as philosophical debates about free will versus determinism as the sources of individual differences in human behavior. Neither Fascist nor Communist states have a monopoly on distorting the science of genetics for their own ends. Such digressions need not detain us here, but numerous references in the bibliography will be of value to those with an interest in the polemics of science.

In this and the next several chapters, we shall attempt to show that genetic factors are essential as a *diathesis* (predisposition) but that they are not sufficient, by themselves, for the development of schizophrenia. Within a broad framework known as *diathesis-stressor* theory, the vast array of facts gathered by both those scientist-clinicians typecast as hereditarians and those typecast as environmentalists can be reconciled without appeasement on either side. At a landmark conference organized by David Rosenthal and Seymour Kety and held at Dorado Beach, Puerto Rico, in 1967, the sage Harvard psychiatrist, Leon Eisenberg, mediating the unique confrontation between the world-class players who preferred biological/genetic explanations and those who preferred experiential/family explanations for the transmission of schizophrenia, concluded, "These findings [adoption and twin studies] persuade me — and I trust most of you — not only that there is a genetic component in the transmission of schizophrenia (a position no more controversial than the defense of motherhood) but that it is a *significant* determinant of the occurrence of the disease" (1968, p. 404). Elsewhere, he speaks of the dangers of both a *mindless* (i.e., ignoring higher order psychological functions) and of a *brainless* (i.e., ignoring biological functions) agenda for studying mental illness (1988). Demonstrating a significant role for genetic factors in the etiology of schizophrenia conveys *no* implications of therapeutic nihilism, either psychotherapeutic or pharmacotherapeutic (Meehl, 1972).

Largely as a consequence of the Dorado Beach conference, the entire field of schizophreniology was converted, at least in public pronouncements, to some kind of interactionist stance for advancing against the common enemy — ignorance about the true causes of schizophrenia.

The development of all human characteristics requires contributions from both genes and environments; in this sense, all characteristics are "acquired" rather than simply inherited or written on a blank slate. For schizophrenia, the unresolved problems are to identify the specific genes, together with their importance as causal factors in the disease, their chemical products, and their mechanisms for modulating behaviors, and to identify the specific environments (physical and psychosocial), together with their contributions to the disease and their mechanisms for modulating behaviors.

The basic evidence for the importance of genetic factors in the etiology of schizophrenia comes from the simultaneous considerations of the results of family, twin, and adoption studies conducted from 1916 to 1989. We shall show how familiality—the tendency of schizophrenia to run in families—once demonstrated, can be interpreted as due largely to the sharing of genes rather than to the sharing of environments and experiences. Our thesis is that schizophrenia is the same kind of common genetic disorder as coronary heart disease, mental retardation, or diabetes, and thus is not like such rare genetic diseases as Huntington's disease (HD) or phenylketonuria (PKU), a form of mental retardation, each with clear, simple Mendelian inheritance patterns associated with dominant (HD) or recessive (PKU) modes of inheritance from parents to children. To understand the basis for this view, it's worth pausing for a moment to be more specific about what we mean when we talk about the genetics of a disorder.

A Genetics Primer

The word *genetics* is used in connection with a vast range of scientific undertakings, from the molecular study of genes and their mutations to the macrolevel study of differences among species and the course of evolution. The field of genetics has very quickly followed the earlier path of the field of physics in dividing the objects of interest into ever smaller particles for analysis. In this book, we usually describe phenomena at the clinical genetics level, technically a part of population genetics; that is, we focus on whole humans and their observable characteristics, or *phenotypes*, and the degree of familiality of those characteristics. We reserve the quarks of genetics for our eager colleagues in molecular genetics. Our knowledge about the specifically genetic characteristics, or *genotypes*, of individuals derives indirectly from our knowledge about the degree of genetic similarity between different pairs of relatives.

The essential paradigm for conducting psychiatric genetics research of the nonmolecular type is implicit in the concept of a human pedigree, or family tree, an invention of Francis Galton. Figure 7 depicts a pedigree,

Degrees of Relationship

		Identical	First	Second	Third
Generation	−3				Great-grandparent
	−2			Grandparent	Great-uncle
	−1		Parent	Uncle	Half uncle
	Current	Proband / Identical twin	Full sibling / Fraternal twin	Half sibling	First cousin
	+1		Child	Nephew	Half nephew
	+2			Grandchild	Grandnephew
	+3				Great-grandchild
Genetic correlation		1.0	0.5	0.25	0.125
Percentage of genes in common with proband		100%	50%	25%	12.5%

FIGURE 7. A scheme for depicting a human family pedigree to show the genetic and generational (environmental) overlap among kinds of relatives for three generations before and after the current one.

but in a form unfamiliar to most people. From the diagram, you can select individuals with the same degree of gene similarity to a benchmark case (termed the index case or proband), but with different degrees of generational (read *environmental*) similarity—for example, parents and offspring. Or, you can select those with different degrees of gene similarity and the same degree of generational similarity—full siblings and half siblings, for example. Once we know the extent of genetic similarity among relatives, we can carry out the quasi experiments of macropsychiatric genetics: The *independent variable*, under the control of the experimenter, becomes degree of gene overlap (from 100 percent for pairs of identical twins to only 12.5 percent for pairs of first cousins), or heredity; and the *dependent variable* becomes the behavior of interest — schizophrenia.

Think about some of the combinations possible and their implications. Similar rates in identical and fraternal twins of schizophrenics, sharing contemporaneous environments but differing greatly in gene overlap, would be evidence of some kind of simple environmental explanation, we observe such a pattern for measles and for language accents. Different rates in siblings and half siblings of schizophrenics reared together, despite sharing environments, would be evidence of some kind of genetic explanation, since only their gene overlap differs, 50 percent versus 25 percent. Similar rates in the siblings and the children of schizophrenics, despite generational differences, would implicate their 50 percent commonality of genotype.

From Simple to Complex Schemes of Inheritance

Simple Mendelian patterns of inheritance—which completely explained the color of flowers for Mendel's pea plants and so exquisitely explain the 50 percent risks seen in the offspring and siblings of HD patients and the 25 percent risks seen in the siblings of PKU patients, together with the 0 percent incidence in their parents—have already been introduced. It is common, regrettably, to hear intelligent people conclude that schizophrenics who do not have similarly affected parents or siblings must have a "nongenetic" form of the disease. They have erred because their knowledge about genetic inheritance does not extend to the more complex and subtle aspects of the genetics of continuously distributed traits and of common diseases. True enough, Victor McKusick's encyclopedic handbook of Mendelian disorders in humans (1988) describes 4,344 Mendelizing phenotypes. Such a view of the genetic disorders afflicting our species, important as it is, omits consideration of the vast numbers of people suffering from the much more common disorders, multifactorial to be sure, in which a significant genetic contribution has been implicated (Rotter, King, and Motulsky, 1990), such as hypertension, diabetes, mental retardation, and coronary heart disease.

With the simple, rare (1 in 20,000 persons affected) dominant gene transmission pattern in Huntington's disease, it could be observed clinically that one parent was always affected (if they lived long enough and there was no doubt about paternity). Furthermore, half of all siblings and half of all offspring, accumulated across families, were eventually affected; and one grandparent of an HD victim would have been similarly affected and he or she would be a parent to the affected parent. Such a classical pedigree pinpointed the mode or mechanism of gene transmission many decades before 1984, when the specific gene was localized to a position on chromosome 4. In 1990, it still is not known what neurochemistry and pathophysiology are involved that lead inexorably to the dementia, and often to the psychoses, seen in HD. Like schizophrenia,

HD has a very wide range in the age of onset; although it usually manifests itself in the 40s, it can appear anywhere between childhood and old age, followed by death in about 15 years. Unlike schizophrenia, gene tests can now be performed that tell, with very high probabilities, exactly which infants born to HD parents will eventually develop the disease — Pandora's box has been opened for the first time for a neuro-psychiatric disease.

Less obvious is the transmission pattern for a simple recessive disease such as PKU. Suppose both parents are carriers of one of the recessive PKU genes. Neither parent has the phenotype — the observable clinical symptoms — but if enough siblings of cases are observed, one in four of them will have the disease. Both parents of cases and half the patients' siblings are genetically different at the PKU chromosome locus (location), but in a silent and invisible manner — they are carriers, in single dose, of the PKU recessive gene. This gene, when present in double dose (*homozygous*, as contrasted with their *heterozygous* parents), leads to a missing essential enzyme for processing the amino acid phenylalanine. Knowledge about the genetics of PKU since its discovery in Norway in 1934 permits homozygous infants with this inborn error of metabolism to be detected by legally mandated blood tests and to be placed on a special diet to prevent the severe mental retardation that formerly was their fate; permits detection of herterozygous carriers so they can receive genetic counseling; and permits the prenatal detection of affected fetuses in families already known to have an affected child. For rare, Mendelian diseases such as HD and PKU, routine pedigree studies are sufficient to pinpoint the mode of genetic transmission.

Much more complex patterns of familial aggregation of pathology have been observed for diabetes, moderate mental retardation, Alzheimer's disease, coronary heart disease, manic-depression (bipolar and unipolar affective disorder), epilepsy, and cleft lip and palate, to name a few. The patterns observed for schizophrenia strike the observer as being in the same ballpark as the conditions just mentioned; for none of these do we find large numbers of families in which the risks to first-degree relatives appear to be close to the expected rates of 50 percent for Mendelian dominant disorders and 25 percent for recessive diseases. Furthermore, a patient with such a *multifactorial-polygenic* condition requiring an accumulation of more than two relevant genes, as well as relevant environmental contributors, most often comes from a family in which neither parent manifests the condition and, usually, none of the siblings or children do either. Paradoxically, these are some of the very reasons to suspect polygenic inheritance as a root cause of such "atypically" familial disorders.

Some of the other hallmarks of polygenic abnormalities are as follows: (1) they are not rare in the general population, occurring with a lifetime risk of greater than 1 in 500; (2) there are gradations of severity ranging from mild to severe, whereas Mendelian diseases are either present or absent; (3) severely ill patients often have more relatives affected than do mildly ill ones; (4) the risk to the next generation increases as a function of the total number of sick relatives in the pedigree — for example, an affected father whose wife's sister is also affected conveys a higher risk to his children than does his affected brother whose wife's relatives are all healthy; (5) the risk drops sharply rather than by 50 percent steps as one proceeds by degrees from very close genetic relatives to more remote ones (48 percent in identical cotwins of schizophrenics to 13 percent in offspring to 2 percent in first cousins); and (6) affected cases appear on both maternal and paternal sides of the family pedigree. A further difference between the Mendelian inheritance of rare disorders and the multifactorial-polygenic inheritance of common disorders is that only in the former do we observe and expect 100 percent of identical cotwins to be affected and, given the mating between two affected parents, do we observe and expect 100 percent (say, in the offspring of two PKU parents) of their children or 75 percent (say, in the offspring of two HD parents) of their children to be similarly affected.

The vast majority of gene-influenced human variation — both normal and abnormal — arises from polygenic effects that, when plotted on paper, take the shape of a bell-shaped, or "normal," curve. Height, IQ test scores, and blood pressure values are nice textbook examples. The numerous polygenes associated with a particular disease coact with each other and other factors to produce pathology, although they themselves are not yet individually detectable. These polygenes are not different in kind from the major genes that cause Mendelian conditions, but each has only a small effect on trait variation as compared to the total variation for that trait. Therefore, the expression of the trait depends much less on which polygenes in the specific system a person has (e.g., height, blood pressure, IQ) than on the total number pulling him or her toward an extreme. A feature of special interest in the study of schizophrenia and other major mental disorders is the ability of such polygenic systems to store and conceal genetic contributors to *the liability to developing* the disorder, somewhat analogous to carrier status for recessive diseases.

For example, let's look at the simplest version of a polygenic system — say, a hypothetical one for height — determined by two locations (*loci*) on the chromosomes and by only two possible genes, **A** or **a** at one locus and **B** or **b** at the other. In the entire population of humans, we would thus expect to find such combinations as **AABB** and **aabb** at the extreme values for height. If we designate either capital letter as height enhancing and either lowercase one as height reducing, those will all capital genes would be the tallest people and those with all lowercase genes would be the shortest people. If both your parents were of average height with genotypes **AaBb**, they would transmit genetically to each of their offspring either an **A** or an **a** and either a **B** or a **b**; such average parents could produce five different height categories of offspring, with a total range of four to zero capital letter genes — **AABB, AABb, AAbb, Aabb,** and **aabb** would be examples of some of the combinations generated by parents of average height. By simply allowing for additional variation in height from prenatal and postnatal nutritional factors, the polygenic model for height becomes a multifactorial-polygenic one.

These kinds of polygenic systems allow, for example, two tall parents to produce offspring shorter than either parent; two parents with average intellectual ability (IQ 100 or so) to have a mentally retarded child (IQ 55 or so) or a genius (IQ 145 or so); and two parents who are not schizophrenic or even markedly eccentric themselves to produce a child who grows up to be a bona fide schizophrenic. Polygenic effects are admittedly elusive when dealing with human beings, but they are routine and palpable when working with plants and animals — as demonstrated by Sewall Wright for extra toes in guinea pigs and Eric Lander and colleagues for variation in tomatoes. The expression of schizophrenia-related polygenic effects at the phenotype level depends on a person's liability to developing schizophrenia, a liability subject to dynamic fluctuations upward or downward depending on the many factors illustrated schematically in Figure 8.

The numbers and kinds of genes inherited in the schizophrenia system are obviously important, but so are hypothetical "antischizophrenia" genes and environments, shown in the figure as genetic and environmental *assets*. We have, in effect, a balance sheet of genetic and environmental assets and liabilities, and whether we in fact develop the disorder or are merely on the edge of "bankruptcy" depends on the bottom line when the assets and the liabilities are summed. The contributors to liability on the five axes of Figure 8 are shown with different weights to convey the idea of different values of the "currency of liability"; some will be like 100-dollar or 50-dollar bills and others will be like fives or ones. Only those individuals in the population whose bottom-line value exceeds a certain *threshold value* at a certain point in time are clinically diagnosable as having schizophrenia.

For coronary heart disease, mental retardation, schizophrenia, and other relatively common kinds of pathologies, the system-specific genes do not act alone. Rather, they coact with other general background genes

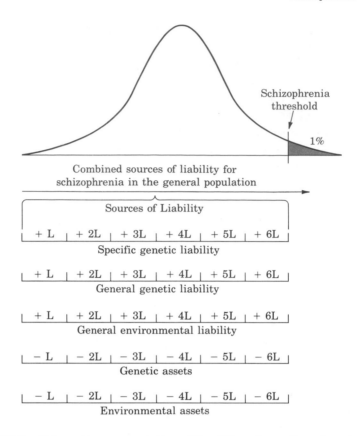

FIGURE 8. Schematic representation of the different genetic and environmental sources of the hypothetical liability for developing schizophrenia on the multifactorial threshold model; the sources take different "currency" values for assets and liabilities and sum, at a point in time, to a value above the threshold shown at the right of the graph (affected with schizophrenia) or below the threshold (well or recovered from schizophrenia).

and with prenatal and postnatal environmental factors. Otherwise, we would not be able to explain the important fact, to be discussed in Chapter 6, that identical twins, despite having all their genes in common, are discordant for schizophrenia half the time (that is, one has it and the other doesn't), but each transmits the liability to schizophrenia to his or her offspring with equal force.

The research detectives who study coronary heart disease (CHD) have already provided exemplars of the kinds of results to expect, by analogy, for a multifactorial-polygenic-threshold disorder (Vogel and Motulsky, 1986). CHD aggregates in families of probands, with rates two to six times higher than in control families; the rates are even higher when the onset is before age 55. (Note that a single-major-locus disease such as HD aggregates in families at 10,000 times the population rate.) Identical cotwins of cases are significantly more often affected than are fraternal cotwins. Numerous contributory risk factors have been identified or implicated, each with its own pattern of weights for genetic and environmental factors. Hypertension, diabetes, and hyperlipidemia (too much fat in the blood) are major risk factors for CHD, and each has an important genetic component. One type of hyperlipidemia involves too much cholesterol due to a unique dominant gene with a prevalence of 1 in 500 people, but the more common type is polygenic with a prevalence of 25 in 500, or 5 percent. Other risk factors are environmental and include obesity, a "couch potato" lifestyle, smoking, a hard-driving personality, alcohol abuse, and living in industrialized societies. DNA markers associated with the proteins, enzymes, and cell-surface receptors for metabolizing lipids have launched a new science of CHD prediction and prevention even in the face of the complications caused by the interaction of the genetic and environmental risk factors mentioned above.

For polygenic characteristics, what is inherited is a *predisposition* toward developing the disorder — a loading of nature's dice that increases the risk of, in this instance, developing schizophrenia. We believe that a relative of a schizophrenic inherits a combined genetic part of the liability for schizophrenia that is much greater (for a son or daughter) or a little greater (for a first cousin) than that for a member of the general population who is genetically unrelated to a schizophrenic. With the increases in liability go increases in risk that exceed the ordinary population risk of about 1 percent. Given the laws of genetics, we can illustrate the overlapping distributions for liabilities in Figure 9. The displacement to the right, toward the threshold for schizophrenia, of the average general liability for first-degree relatives (siblings and children) has the effect of "pushing" many more of them over the threshold value of liability that leads to overt schizophrenia; the distribution curve for third-degree relatives (first cousins) of schizophrenics is displaced much less to the right as a consequence of a diminished amount of genetic overlap (12.5 percent versus 50 percent for first-degree relatives).

By combining the concepts in Figures 7, 8, and 9, you can better appreciate that each kind of relative, in the individual case, has a wide range of possibilities in a complex game of gene-environment Russian roulette; only the averages behave predictably. The combined liability can be modified by many factors, it can be suppressed completely from expressing itself as schizophrenia, or it can be made much worse. That is the essence of the diathesis-stressor formulation — Nature proposes and Nurture disposes.

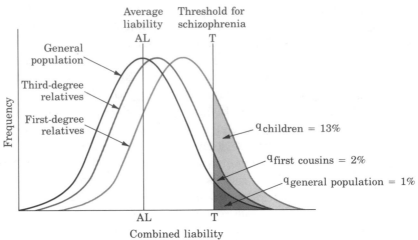

FIGURE 9.   A schematic model depicting the polygenic inheritance of threshold characteristics: Three distributions of the underlying liability for schizophrenia in the general population with a risk of 1%, in the third-degree relatives (cousins) of schizophrenics with a risk of 2%, and in the first-degree relatives (siblings and children) with a risk of 13%. T = threshold above which schizophrenia is clinically observed; AL = average liability in the general population. Note that many relatives of schizophrenics have liability values *less* than those of members of the general population who are unrelated to schizophrenics.

The strength of a diathesis-stressor framework that incorporates a polygenic basis for the diathesis is that it fits the results of many careful studies. However, a single-major-locus genetic hypothesis (a competing genetic model) may be correct for a few cases of schizophrenia; a certain amount of etiological heterogeneity is to be expected with multifactorial disorders, as illustrated with CHD. Such exceptions will be important to detect because their treatment and the counseling advice may well be quite different.

A plurality of data-informed scientists who do research on schizophrenia accept the diathesis-stressor formulation or something close to it. The previous statement was worded so as to exclude marxist philosophers, orthodox psychoanalysts, and assorted ideologues, whether politically motivated or not. They would rather grind their own axes than further the impartial quest for the causes of schizophrenia. Among scientists concentrating on the root causes of schizophrenia, the remaining disagreements are about the best currency values or "weights" to give the contributors to liability shown in Figure 8.

No purely environmental theory has been borne out by any study. No one has ever proved that environmental factors alone can be a sufficient cause of true schizophrenia in any person not related to a schizophrenic. Schizophrenia has been diagnosed after, for instance, a head injury or the flu, but these trauma-induced, schizophrenia-like psychoses are no longer accepted as true schizophrenia; the majority are now seen as parts of the broader category of organic brain syndromes associated with physical disorders. An ambiguous "no-man's land" still exists for those few cases of schizophrenia triggered by a trauma that is just one more risk-increasing factor in the combined liability. A synthesis of competing explanations for the mode of transmission of schizophrenia will be postponed to Chapter 11, after all the relevant data have been sampled.

Foundations for a Genetic Position

Our case for the role of genetic inheritance as a major source of schizophrenia is circumstantial, but may well be sufficient to "convict the culprit." It is based on clinical population genetics and it implies, but does not prove, that there is a neurochemical and/or neuroanatomical cause for the pattern of behavioral aberration. While research on the biology of schizophrenia goes on, scientists continue to study schizophrenia to uncover the mechanisms of inheritance with clinical analyses and with DNA probes, to look for environmental influences, and to predict outcomes or levels of risk. Early research going back to 1916 laid a foundation for our knowledge; today's research continues to build understanding.

Family, twin, and adoption studies of schizophrenia provide the grist for our mill. Each contributes to the genetic argument, complementing the others. No one method alone yields conclusive proof or disproof. For some psychiatric conditions, such as bipolar affective disorder (manic-depressive illness), linkage studies with DNA probes within a few rare families heavily loaded with one or two forms of the disorder suggested that a dominant gene on chromosome 11 in some Pennsylvania Amish families and a different dominant gene on the X chromosome in some Israeli families cause bipolar disorder. The Amish-based claims about chromosome 11 reported in 1988 were withdrawn as nonreplicable in 1989 after diagnostic criteria were refined and new cases fell ill. If such linkage results do get replicated, they will shed enormous light on the genetic aspects of mental illness, because they will turn the circumstantial evidence of genetic involvement gained from population genetic strategies into hard, physical evidence. For schizophrenia, devoting too many resources to a similar kind of random, long-shot search for genetic markers, absent some candidate genetic marker that is logically related to the neurochemical and pathophysiological facts established by research, would be like investing in penny stocks on the Denver Stock Exchange. However, once in awhile someone makes a fortune by just such speculation. Preliminary suggestions about a gene on chromosome 5 that was implicated by linkage analysis and that may be involved in the schizophrenias seen in a few English and Icelandic families will be discussed in Chapter 12.

Family Studies

If transmissible genetic factors contribute to developing schizophrenia, the disorder should cluster in affected families at a higher rate than the 1 percent affected in the general population. That a disease is familial, however, does not necessarily mean that it is genetic. As we have already cautioned, it may be transmitted through families not by genes but through some cultural practice, some infectious source, or some kind of learning by imitation. When we look at family studies to decide whether the variable is something genetic, we want to ask if the pattern of risks rises and falls as a function of the genetic overlap rather than as a consequence of shared experience. Do closer relatives, who have more genes in common, have a higher risk, and does that risk diminish as family relatedness diminishes? Furthermore, if the relatives are not overtly affected with schizophrenia, do they have something specifically odd or different about them that might provide clues to their having inherited a "diluted" form of schizophrenia; that is, schizophrenia spectrum disorder?

The first systematic family studies of schizophrenia were done in 1916 by Ernst Rüdin working with Kraepelin in Munich, as noted in a previous chapter. In 1932, Bruno Schulz, a star member of Rüdin's Munich school (now the Max Planck Institute for Psychiatry), elected to follow up Rüdin's families. He found that 8.1 percent of the more than 2,000 sisters and brothers of the 613 original subjects with unaffected parents were schizophrenic. Using the classical Kraepelinian subjects of schizophrenia to examine the possibility of qualitative differences in the inheritance of schizophrenia, Schulz observed that there was a strong tendency for familial "homotypy." For instance, if the original subject was a catatonic schizophrenic, other affected siblings tended to be catatonic subtypes also, although different subtypes did frequently occur within the same family. Schulz showed, however, that no matter what subtype the original subject was, his or her brothers and sisters had a high risk for some type of schizophrenia. Such results argue against dividing up schizophrenia into pre-Kraepelinian pieces and for the view that differences in the symptom picture could be accounted for by a dimension of severity in *one* disorder.

Schulz next anticipated modern diathesis-stressor formulations by subdividing his original subjects according to whether there was an external life event — physical or psychological — that precipitated the illness, or whether it appeared out of the blue. As a result, he was the first researcher to say that schizophrenias that follow a somatic stressor, such as a head injury, often may not be genetically the same as those in which no such precipitant occurred. He found that risk in *siblings* of cases with no known "trigger" was as high as 10 percent; for his sample, this was twice the 4.8 percent found in somatically triggered cases, implying greater genetic diathesis or "loading" in those cases with little evidence of external stressors. Later, in a variation of this idea, Schulz looked at the relatives of unrecovered and recovered schizophrenics in the days before there were specific treatments for the disorder; the siblings of the latter group had a risk for schizophrenia of only 3.3 percent, and the children a risk of 7.4 percent (cf. Figure 10 for an average risk of 13 percent).

Franz J. Kallmann, progenitor of psychiatric genetics in the United States, published the first comprehensive study of all branches of schizophrenics' pedigrees, including a detailed, major study on the children of schizophrenics, in 1938 shortly after he had emigrated as a refugee from Hitler's Germany to work at the New York State Psychiatric Institute. He studied the offspring of 1,087 schizophrenics admitted to a Berlin psychiatric hospital at the turn of the century. Altogether, Kallmann gathered hospital-chart data on 14,000 persons in five generations, two generations on each side of the turn-of-the-century schizophrenic index cases. The risk figures for schizophrenia that he obtained from each class of relatives — parents, grandparents, aunts, etc. — contribute importantly to the summary data in Figure 10. One of the additional conclusions that can be derived from Kallmann's results is that the severity of the illness of the sick parent is a factor in the schizophrenia risk to his or her children, thereby supporting the earlier work by Schulz. He showed that the more severely ill the parent was (he called these severe cases *nuclear* schizophrenics; Kraepelin had called them hebephrenic and catatonic), the more likely his or her children and siblings were to be schizophrenic; the offspring of the nuclear cases had a 20 percent risk, while those of the paranoid and simple schizophrenics had a risk of only 10 percent. As in the Schulz studies, all types of schizophrenia appeared in all types of relatives; from such results, we can conclude that a dimension of severity, or "gene dosage effects," is operating rather than that

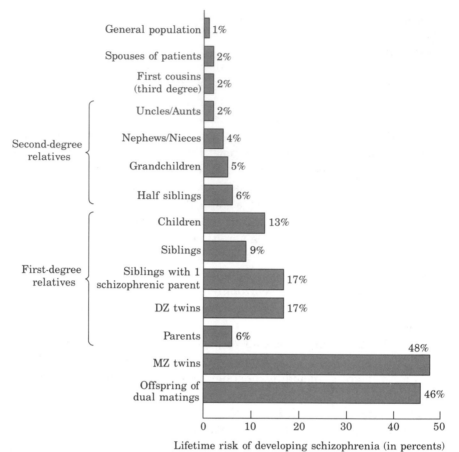

FIGURE 10.   Grand average risks for developing schizophrenia compiled from the family and twin studies conducted in European populations between 1920 and 1987; the degree of risk correlates highly with the degree of genetic relatedness.

this is an indicator of two different diseases — nuclear/severe schizophrenia versus peripheral/mild schizophrenia.

Figure 10 summarizes the lifetime risks of being affected with schizophrenia for the various kinds of relatives of a schizophrenic. The figure shows the degree of genetic relationship for different groupings of relatives which can be compared with the percentage of genes that various family members share in common (cf. Figure 7) with a schizophrenic proband. The data come from pooling information from about 40 reliable studies conducted in western Europe from 1920 to 1987. There is some danger in pooling such information, because the study designs are not exactly alike, but pooling after the judicious removal of the poorest data gives a clear and stable summary pattern not obtainable from any one or two studies.

We have used studies from Germany, Switzerland, the Scandinavian countries, and the United Kingdom in generating Figure 10, because these countries have similarly conservative diagnostic standards. Included are a modern replication of the Kallmann family study completed in Sweden in 1970 by David Kay and Rolf Lindelius and involving 4,000 relatives and a Swiss family study completed by Manfred Bleuler (1978) involving 3,000 relatives; the orientation of these investigators was clearly "interactional" and they had no genetic axe to grind. The European researchers, unlike those in the United States, have better access to relatively stable populations and often can use the national psychiatric registers generated by systems of national health insurance. Such factors, together with relative (compared to the United States) homogeneity for race, religion, and social class and a high degree of cooperativeness from relatives, lead us to emphasize European data for obtaining meaningful pooled results.

Figure 10 presents the risks of developing schizophrenia for first-degree (parents, sibs, children), second-degree (uncles, nephews, grandchildren, half sibs), and third-degree (first cousins) relatives of schizophrenics. It also gives the risks for fraternal twins, who are also first-degree relatives, and for identical twins, who could be called zero-degree relatives because they are really genetic clones of one another. Risks for spouses, who are highly unlikely to be related genetically, are presented to test the idea of being able to induce schizophrenia in someone simply by sharing an intimate physical and psychological relationship. It also includes the risks of schizophrenia for the children of two schizophrenics, showing them to be nowhere near the 100 percent or 75 percent risks mentioned earlier for recessive and dominant genetic disorders. Except for this category, which we'll get back to later, the pooled studies represent large enough samples to be quite reliable. An enormous amount of information is summarized in the figure, and it deserves careful study.

Interpreting the Pattern of Familial Risks

Overall, the pattern of risk figures in the relatives of schizophrenics strongly supports the conclusion that the magnitude of the increased risk varies with the amount of gene sharing and not with the amount of experience sharing. Identical twins and offspring of dual matings have higher risks than do first-degree relatives, who have higher risks than do second-degree relatives, who have higher risks than do third-degree relatives, who have higher risks than do spouses and the basic risk of 1 percent in the general population.

Concern has been raised recently that the earlier overwhelming data showing the familial clustering of schizophrenia could have been falsely generated by unblindfolded, genetically biased psychiatrists using unreliable clinical judgments. Impetus was added to these concerns when two new family studies from the United States with large numbers of first-degree relatives of schizophrenics were evaluated using the "objective" criteria of DSM-III. One study found no schizophrenics among 199 relatives, and the other found 2 among 128 first-degree relatives; based on Figure 10, we would expect about 10 percent from each sample to be affected. Have these modern researchers uncovered a cover-up? Not really. The negative studies had not bothered to reduce the sample sizes by using an appropriate age correction, they did not apply their criteria to a normal control sample to obtain a comparison figure, and they relied too much on secondhand information. Other family and twin studies quickly followed using the *same* objective, criteria-based diagnostic schemes and found rates in first-degree relatives some 9 to 18 times higher than in control groups *evaluated with the same standard*, thus confirming the pattern of familial risks shown in Figure 10.

The risk figures within first-degree relatives (parents, sibs, and children), all of whom have the possibility of a 50 percent gene overlap, is not uniform, and we need to look at what this might mean. Let's start with the small variation between the children of schizophrenics, 13 percent, and the brothers and sisters of schizophrenics, 9 percent. The higher risk in children may well be associated with the exposure to noxious environmental stressors or triggers provided unintentionally and haplessly by schizophrenic parenting; being a sister or a brother to a schizophrenic is likely to be less stressful than being the child of one. However, recent studies have reported equal risks in siblings and in children of schizophrenics. Adoption studies, considered in Chapter 7, are needed to further our understanding here.

Within the class of first-degree relatives, parent-child pairs share *exactly* 50 percent of their genes, but pairs of siblings have *on average* 50 percent of their genes in common. Thus, it is possible for some sibling pairs to have considerably more or fewer than half their genes in common, with obvious consequences for their similarity or dissimilarity in polygenically influenced characteristics (such as height, blood pressure, and the liability for developing schizophrenia).

The observed 6 percent risk to the parents of schizophrenics is only half the risk observed for other first-degree relatives; this is an important fact that appears to conflict with either a dominant gene or a polygenic theory prediction of equal risks in parents, siblings, and children. The vast majority of schizophrenics do not have conspicuously abnormal parents, as anyone attending an annual meeting of the National Alliance for the Mentally Ill will have noted. The best explanations for the lower risk in parents of schizophrenics center about the processes of social selection for marriage and parenthood. Schizophrenics and preschizophrenics are less likely to marry than are nonschizophrenics; those schizophrenics who do marry tend to have the least severe cases of schizophrenia and/or have a later age of onset for their first serious symptoms. A post-World War II Swiss study conducted by Manfred Bleuler (1978) found that of 28 schizophrenic parents of schizophrenics, 23 had their onset after age 30 and 19 had mild cases or very good recoveries from an episode of illness. These and other selective forces explain not only the lower rate of schizophrenia in the parents of schizophrenics compared to the other first-degree relatives, but also the considerably lower rate for parents in general — it is half the usual 1 percent rate observed in the general population unselected for parenthood.

Among schizophrenics who do become parents, the ratio of mothers to fathers is two to one. This striking fact, coupled with the high risk for schizophrenia among the children of schizophrenics, has led some partisans to proclaim confirmation of the notion that pathogenic mothering causes schizophrenia. The term *schizophrenogenic mother* was coined by the psychoanalyst Frieda Fromm-Reichmann in 1948 during the heyday of explanations favoring child-rearing attitudes as the central cause of schizophrenia, with no reference whatsoever to genes. The more neutral and, we believe, valid explanation is that age of onset is earlier in males, thus reducing the likelihood of their finding a mate, and that marriage and procreation occur earlier in females, thus allowing the future female schizophrenic increased opportunity to marry and start a family before she becomes ill. The net effect of such social forces of selection results in twice as many children being born to schizophrenic mothers as to schizophrenic fathers, thus feeding the myth of schizophrenogenic mothering. In the three studies that provide data on the matter, the *risk* of developing schizophrenia in the offspring of schizophrenic mothers is the same as that observed in the offspring of schizophrenic fathers. The phrase "schizophrenic mother" can now be expunged from the scientific literature.

Conventional wisdom says that although only 13 percent of the children of schizophrenics are themselves schizophrenic, many more, perhaps another 50 percent or so, are conspicuously abnormal psychiatrically. If this were true, we would be forced to revise our genetic theory to say that schizophrenia is inherited via a dominant gene, a gene that does not transmit schizophrenia itself, but a schizophrenia-related illness, or to adopt a more psychogenic theory of cultural transmission. Three early German studies on the children of schizophrenics, including the Kallmann one, are the sources of this belief in high rates for combined deviance. We find the data untenable in light of newer studies.

Manfred Bleuler, carrying on his father's work in Switzerland (1978), found a surprisingly high percentage of normal children among the off-spring of the schizophrenics he and his father had studied. The younger Bleuler followed the patients for more than two decades and knew them well; he knew their own histories, the histories of their spouses, and the histories of their marriages. He considered fully 74 percent of the children over age 20 to be completely normal. His own explanation is informative.

If the hitherto prevailing dogma . . . is to be retained, to the effect that over half the children of schizophrenics are in some way abnormal (and therefore basically "undesirable"), it is in serious need of more exacting definitions. These children can only fit that classification if "abnormal," "pathological," or "undesirable" are terms to be applied even to those whose behavior becomes different from and more difficult than that of their basic nature because of an added stressful situation. They will not fit that classification if the normal person is permitted to exhibit behavior traits in stressful situations that might be difficult to distinguish from abnormal or psychopathic behavior patterns. (1978, p. 381)

This quotation makes clear that the kinds of psychopathology observed in the relatives of schizophrenics, especially in the children, can be both genetically transmitted and reactive and even adaptive to the presence of a psychotic, obnoxious, and guilt-provoking parent in one's psychological space. It will take both twin *and* adoption strategies to add perspective to these complementary interpretations.

Exploring Diagnostic Boundaries

Family studies show that neither schizophrenia nor affective disorders always breed true to type. Individuals with each core diagnosis are more likely to have relatives who fall under the same, or homotypic, diagnosis than under the other, but there is some crossover. Schizophrenia and major affective disorders can appear in the same families just by chance, given the lifetime risk of each (1 percent and 8 to 18 percent, respectively). The British psychiatrist Peter McGuffin and his colleagues have described the complexities of diagnosis in a set of identical triplets, two of whom had hospital-chart diagnoses of schizophrenia and one of whom was called manic-depressive, manic type. Three clinicians who did not know that the subjects were related agreed closely with the hospital diagnoses, but the more formal and objective Research Diagnostic Criteria allowed all three to be called schizoaffective psychosis, manic type, at different ages and stages. Upon learning that the subjects were related, the judges agreed that these identical triplets all had variants of the same disorder, but that it was some kind of schizophrenia that did not fit neatly into one of our textbook pigeonholes.

At Super-High-Risk: The Offspring of Two Patients

Despite the rarity of matings between schizophrenic mothers and schizophrenic fathers, there are five completed studies to learn from. The total number of adult offspring in these studies is small, with an age-corrected sample size of only 134 risk lives observed, but the results were similar enough across studies to be taken seriously. Early workers were eager to study these matings because the crosses might reveal Mendelian patterns. If schizophrenia were caused by a dominant gene, 75 percent of the children would be affected. If it were a recessive, 100 percent would be. (And, to anticipate Chapter 7, if it were transmitted simply by exposure to schizophrenic parents or parenting, 100 percent would be.) The results were fairly uniform. About one-third of the children of two schizophrenic parents were schizophrenic. With appropriate correction for age, this yields a lifetime maximum risk of 46 percent. The studies also showed, however, that the children of two schizophrenics were at considerably higher risk for other psychiatric abnormalities. The largest percentage were neurotic, rather than psychotic, but some of them (perhaps as many as one-fourth) were normal. This final result is amazing, considering how much instability there must be in lives of children who have two schizophrenic parents. How can upbringing in a household loaded with stress be a major factor contributing to schizophrenia if one-fourth of the children who have problem genes from both sides of their inheritance and who live in a family disrupted by two crazy parents turn out to be normal? What a tribute to the resilience of the human body and spirit — and good genetic luck! The interaction of stress *plus* specific genetic vulnerability could produce such results.

Conducting studies of families in which both parents are schizophrenic is difficult, because even in a large population, such matings are rare or childless. When we do find such couples, they and their offspring provide extremely valuable benchmark information about the maximum possible genetic plus psychogenic risks. Recent statistics gathered in Scandinavian studies indicate that some interesting information may emerge from research using psychiatrically hospitalized parents of several types —

reactive psychosis, severe personality disorders, schizophrenia, and affective disorder. We hope to obtain the kind of genetic information that surprised scientists studying deafness half a century ago. When the opening of schools for the deaf began to foster more marriages among deaf persons, new information came to light. Scientists had believed that most forms of congenital deafness were recessive and a few were dominant. But these marriages turned up a wide variety of offspring with respect to deafness. In some cases where both parents appeared to have the same recessive form of deafness, *none* of their offspring were deaf, although all were expected to be. Similar findings emerged from studies of genetic blindness.

Gottesman and Aksel Bertelsen, a Danish psychiatrist working in the Institute for Psychiatric Demography, have been following up the 378 adult offspring of the 139 couples in the Danish National Psychiatric Register who had both been hospitalized for some kind of severe mental disorder prior to 1961; all subjects were followed to the end of 1983. Initial results show that of the 30 schizophrenic offspring produced, 19 had a schizophrenic mother or father and 8 had a manic-depressive parent but not a schizophrenic one; only 3 (0.8 percent, the general population risk) resulted from other mating patterns. Fully 53 percent of the offspring of these 139 super-high-risk matings were not conspicuously abnormal. Another interesting finding was that a diagnosis of reactive or psychogenic psychosis (a third type of psychosis, neither schizophrenic nor affective, and clearly precipitated by stressful events) in a parent did not lead to any increased risk for mental illness among the offspring; the result implies a category of psychosis that is wholly environmental in origin.

Summary

It was not until the beginning of the twentieth century that genetic strategies could be incorporated into the search for the causes of psychiatric afflictions such as schizophrenia. Carefully collected data conclusively establish that the disorder is familial, with risks in relatives many times greater than the population risk of 1 percent, and with the risks paralleling the proportion of genes shared with a sick relative. Familiality is necessary to prove the genetic argument, but it is certainly not sufficient, as noted a number of times in this book. Let us remind ourselves of some very important facts. The vast majority of schizophrenics will have *neither* parent who is overtly schizophrenic — some 89 percent — and will have *neither* parents nor siblings who are affected—some 81 percent. Furthermore, a sizable majority—about 63 percent—will have *negative* family histories—that is, "clean pedigrees"—even allowing for such first-degree relatives as children and such second-degree relatives as nieces and nephews. Risks for schizophrenia in such distant relatives as grandparents, aunts, uncles, and cousins provide little useful information, because they are too close to the general population risks to permit the detection of familial aggregation. Twin and adoption studies of schizophrenia are still needed to further challenge the possibility that the high rates seen in relatives are a consequence of shared experiences with a schizophrenic, and they will be examined in detail in the next two chapters.

6

The Evidence from Twin Studies

Although the family studies reviewed in the last chapter show that the relatives of schizophrenics are more likely to be schizophrenic than is the general population, these studies only *implicate* genes as the source of this familial illness. It could still be argued that the excess risk is caused by shared family environment and experience or psychological contagion. The classical studies of identical (monozygotic, MZ, or one-egg) and fraternal (dizygotic, DZ, or two-egg) twins reared together test the idea that genes are the cause of the familiality. The strategy derives from the fact that identical twins share 100 percent of their genes and their environment, whereas fraternal twins share, on average, 50 percent of their genes and 100 percent of their environment. In most studies, same-sex fraternal twins are used to eliminate possible variations due to sex differences within a pair. If shared family environment is of primary importance in causing schizophrenia, then all cotwins, regardless of zygosity, would be similar. Identical pairs would be no more alike (no more concordant for schizophrenia) than would fraternal pairs. If genes are more important, identical twins would be considerably more concordant for schizophrenia than would fraternal twins.

Sir Francis Galton, the British gentleman-scientist and cousin of Charles Darwin mentioned earlier, is also considered to be the first to use twins to study the "relative powers of nature and nurture" (1875). His ideas were intuitive rather than scientifically rigorous, but he did describe a pair of identical twin brothers concordant for some kind of psychosis, then termed "monomania," whose symptoms included paranoid delusions, auditory hallucinations, and mood swings — it could well have been schizophrenia. The twin method was not developed scientifically until 1924, by the German scientist H. W. Siemens. This method has since become one of the basic strategies for gathering initial data relevant to the possible role of genetic factors for conditions of uncertain etiology. Given that 2 in every 100 adults is a twin, samples of healthy twins can be collected readily, and, depending on the population risk for a particular disorder or condition, samples of affected twins can be collected with some special efforts. In the United States and Europe, about 33 pairs of identical twins and 70 to 90 pairs of fraternal twins are born per 10,000 deliveries; older mothers have more fraternal twins, but age is not correlated with MZ twinning.

When conducting research, twins can be separated routinely into MZ and same-sex DZ pairs by a combination of blood typing and observations of general similarity in appearance (eye color, hair color and texture, height, weight), sometimes including fingerprint analyses. Detailed elaborations of the twin method can be found in Bulmer (1970), Smith (1974), and Gottesman and Carey (1983). It is sufficient here simply to appreciate the conceptual simplicity of the method:

Identical Twin Similarity for Schizophrenia (is due to) = 100 % of genetic variance + 100% of environmental variance

Fraternal Twin Similarity for Schizophrenia (is due to) = 50% of genetic variance + 100% of environmental variance

Therefore: Identicals minus Fraternals Similarity = 50% of the genetic variance in the liability to developing schizophrenia

Classical Twin Studies: 1928–1953

Altogether, the world research literature on schizophrenia contains only a dozen systematic studies, each labor intensive and each valuable for the insight it gives into the complexities of the contributors to the causes of this psychosis. Anecdotal reports of mental illness in twins, usually identical and usually concordant for the disease in question, have spiced the medical and behavioral literature since the days of Galton. Lacking systematic data on both fraternal twins, as control for the factor of shared environments, *and* identical twins, such stories had little scientific merit. In regard to measles, for example, almost all identical twin pairs living together in childhood will be concordant; by noting that the concordance rate for fraternals is also close to 100 percent, hypotheses favoring a gene for measles can be ruled out. For a behavioral characteristic such as juvenile delinquency, MZ and DZ concordance rates, 87 percent versus 72 percent, are too close to give much weight to genetic factors. For adult criminality, however, the extent of the difference between the MZ rate of 51 percent and the DZ rate of 22 percent does implicate a significant role for some kind of genetic contribution (see McGuffin and Gottesman, 1985).

When the first systematic study of twins was conducted by Hans Luxenburger in 1928 at Rüdin's Munich Institute, a genetic cause of schizophrenia was already taken for granted, and the goal was to calculate the "penetrance" of the alleged gene that caused it, because results from risks to siblings did not fit the neat 25 percent or 50 percent expectation of a Mendelian theory. (*Penetrance* is a convenient concept invoked after the fact to *explain* the failure of the data to behave the way textbook rules say they should; with human data, as opposed to data on plants and bacteria, the term betrays our ignorance about what is happening at the basic levels of gene action.) Luxenburger was disappointed with the results he obtained. They were not as conclusive as he had hoped, and even when both identical twins in a pair were schizophrenic, they did not show the "photographic similarity" (similar onset, similar severity, and so on) he had expected from case studies reported in the literature. Nevertheless, both identical twins clearly were affected with schizophrenia more often than were both fraternal cotwins. Best estimates of his results are that 58 percent of identical twins (11 of 19) and none of 13 fraternal cotwins of schizophrenics were themselves affected. Once again, too small a sample generates more noise than signal: The zero rate in fraternals implies, by itself, *neither* a genetic nor an environmental influence.

Like most twin studies, the sample examined by Luxenburger was small even though it was the total yield from screening 16,000 psychiatric patients against local birth records. In the general population, schizophrenic twins are expected only as often as the lifetime risk of schizophrenia in adults (1 percent) multiplied by the prevalence of twins in the

(U.S. white or European) population (2 percent). The formula is $0.01 \times 0.02 = 0.0002$; so 2 in 10,000 persons will be both schizophrenic and a twin, but only 1 in 15,000 will be an identical twin with schizophrenia. Small sample sizes are the rule in twin studies of psychopathology.

In 1934, Aaron J. Rosanoff published a U.S. and Canadian twin-record study that is impossible to evaluate because of various methodological problems (e.g., no personal interviews were conducted), but the findings were surprisingly similar to those of the other studies (see Figure 10, p. 96). He concluded that 61 percent of 41 identical (MZ) pairs and 13 percent of 53 same-sex, fraternal (SS DZ) pairs were concordant for schizophrenia.

Erik Essen-Möller's 1941 study from Sweden with a follow-up in 1971 is much more reliable and easier to follow than Rosanoff's because Essen-Möller provided excellent case histories. As the most eminent Swedish psychiatrist of the last half century, informed by postdoctoral experience at the Munich Institute, his monograph deserves close study. He based his data on consecutive admissions to Swedish mental hospitals, and he verified twinships from birth registers. Yet in considering Essen-Möller's data, we must remember that he is a strict, conservative diagnostician. When he counted major mental illnesses with schizophrenic features as concordant, rather than only his definite schizophrenics, his rates are 64 percent for 11 MZ pairs and 15 percent for 27 SS DZ pairs. His views on character traits associated with schizophrenia, such as *schizoidia*, derived from his scrutiny of the personalities of cotwins, provide an important foundation for contemporary ideas about schizophrenia spectrum conditions.

By far the best known of the early twin studies was published by Kallmann in 1946, based on research conducted in the United States and influenced by his huge German family study (see Chapter 5). His militant, hereditarian point of view has prevented a full appreciation of his data. He sampled the resident population of all state hospitals in New York in 1937 and subsequent admissions to those hospitals up to 1945 and located 174 MZ and 517 DZ pairs with a schizophrenic member. Even though Kallmann's is the largest and best known study, it has also been the most criticized for exaggerating the role of genetic factors in schizophrenia. Some, but not all, of the criticism is justified (Shields, Gottesman, and Slater, 1967). The study is basically sound, but it is sadly lacking in details. At the end of his study, after some diagnostic revision, he found 59 percent MZ and 9 percent DZ twin pairs concordant for what he called "definite schizophrenia." When his doubtful schizophrenia cases were included, his concordance rates increased to 69 percent and 10 percent. These values are very high, but get even higher with routine age correction, to 85 percent and 15 percent, making both supporters and opponents edgy. Two of the most likely reasons for the unusually high rates are that his sample came from the back wards of state hospitals before the modern treatment era began — it was overweighted with severe, chronic cases with greater genetic and environmental liabilities — and that standard age correcting *over*corrected. It has been observed that identical twins who are going to become concordant for schizophrenia do so within 3 to 5 years after the first twin falls ill. Therefore, Kallmann's sample, which was loaded with older and followed-up cotwins, needed less than the standard amount of age correction applied to it.

Concurrently with Kallmann, starting just before and finishing shortly after World War II, Eliot Slater, a renowned British psychiatrist with postdoctoral training in Munich under Schulz, and James Shields, an eminent behavioral geneticist, carried out a twin study in London covering all mental disorders. Their work was based on the total resident population and subsequent admissions to the numerous London County Council mental hospitals (equivalent to our state hospitals). They broke their study into four categories of psychopathology, including one group of schizophrenics. For these, the concordance rates were 65 percent for 37 MZ pairs and 14 percent for 58 SS DZ pairs. As in the Swedish study, case histories are provided for further reanalysis. Minor deviations in the personalities of non psychotic first-degree relatives of the twins suggested to Slater and Shields that traits such as lack of feeling, suspiciousness, being eccentric, and being exceedingly reserved might provide clues to a genetically influenced and etiologically relevant schizoid personality —evidence for an objective definition of spectrum conditions. Once again, twin studies were shown to be most useful when they were turned into twin-family studies with attention paid both to the cotwins and to the other relatives.

After the 1953 U.K. study, no other twin studies were reported until 1961, when Eiji Inouye announced the results of his study in Japan. No one else has reported a nonwhite non-Western schizophrenic twin sample. Inouye's original concordances were revised in 1972, combining categories of schizophrenia and schizophrenia-like psychoses, to 59 percent of 58 MZ pairs and 15 percent of 20 DZ pairs. Inouye noted that concordance in identical twins was much higher when the first twin was severely, chronically or recurrently schizophrenic than when the first affected twin was mildly, chronically or transiently schizophrenic; all unaffected MZ twins were thought to have schizoid personalities. Both observations are more consistent with a quantitative, or polygenic, model than with a Mendelian model.

Contemporary Twin Studies

The six classical twin studies described above showed remarkably similar concordance rates without the dubious corrections for age. These corrections are dubious because they assume no correlation in the age of onset for pairs of relatives but, as noted earlier, the correlation in the age of onset for twin pairs is considerable. Despite the consistency of the twin findings with respect to schizophrenia and their implications for the importance of genetic factors in causing schizophrenia, standard texts in psychopathology ignored or belittled them. It was only at and after the Dorado Beach Conference in 1967 that a renewed appreciation began for twin data as a major piece of evidence. The five twin studies that have been reported since 1963 — three from Scandinavia, one from the United States, and one from the United Kingdom — generally report lower identical twin concordance rates than did the earlier studies, but similar fraternal rates. On the surface, their MZ findings range more widely, actually from zero to 50 percent, but the zero from the preliminary reports of the Finnish study is a fluke. Since virtually all these studies include twins who are still alive and discordant, their results are not yet final. For instance, the Finnish study first reported no concordance among rather young identical twins, but follow-up some years later resulted in a 36 percent concordance. Each time researchers recheck on their original subjects, new rates may result, but changes will be minor after five to ten years following diagnosis in the first twin of identical pairs.

The new studies are critical to sustaining the credibility of the body of genetic information. The classical studies were heavily criticized by, for example, the influential psychodynamic psychiatrist Don Jackson (1960), for being too biased by hereditarian investigators, for contaminating the information on zygosity (twin type) with unblindfolded psychiatric diagnoses by the sole investigator, for the atypicality of twins as subjects because of their alleged "identity confusion," and for not studying identical twins who had been reared apart from each other to exclude imitation as an explanation for concordance. David Rosenthal was a much more reasonable taskmaster and constructive critic in his series of early 1960s papers that provided a blueprint for improving the state of the art of schizophrenia twin studies (see also Shields, 1968).

Table 8 presents a survey of these newer studies. It gives a range (not a single percentage) under the heading *pairwise* concordance, as reported by the investigators themselves, to permit direct comparisons with the classical twin studies. We also have converted that figure to a technically

TABLE 8

Concordance rates for schizophrenia in newer twin studies

Country/year	MZ Pairs			DZ Pairs		
	Total	Pairwise rate (%)	Probandwise rate (%)	Total	Pairwise rate (%)	Probandwise rate (%)
Finland, 1963/1971	17	0–36	35	20	5–14	13
Norway, 1967	55	25–38	45	90	4–10	15
Denmark, 1973	21	24–48	56	41	10–19	27
United Kingdom, 1966/1987	22	40–50	58	33	9–12	15
Weighted average			48%			17%
United States, 1969/1983	164	18	31	277	3	6

more correct and more genetically informative measurement called a *probandwise* concordance. For a pairwise concordance, a pair, both of whom are schizophrenic, is counted as one pair in the numerator and in the denominator. For a probandwise concordance, concordant twins are counted as two pairs both in the numerator and in the denominator, but only when those pairs contain schizophrenics that were both identified from the official register of cases independently. Probandwise rates are unbiased in regard to sampling and are the only rates that are directly comparable to the population risks reported in previous chapters (see Gottesman and Shields, 1976). If we had been able to report probandwise rates for the earlier studies, we would have, but the necessary information was not available.

The first of the modern studies was conducted in Finland by Pekka Tienari (1963, 1975) a psychodynamically and family-systems oriented psychiatrist. It is based on a nationwide birth register of male twins born from 1920 to 1929. Childhood deaths and lengthy wars led to the loss of half his original subjects, but he was able to identify psychoses of all types in 1,000 intact surviving pairs. To everyone's surprise, when he originally reported on his 16 MZ pairs with schizophrenia in 1963, none of them were concordant. Although the sample was small, no one expected zero concordance, because that would imply that neither heredity nor shared environment played a role in causing schizophrenia! Tienari, on follow-up (1975), reported a pairwise concordance rate of 36 percent for MZ twins and 14 percent for DZ twins and is no longer the odd man out.

In 1967, Einar Kringlen, skeptical about received genetic dogma and impressed by the influence of social forces on human behavior, published a two-volume work based on his Norwegian study, including case histories for all psychotic identical twins. He used the country's well-known central register for psychosis and the twin birth registers for the years 1901–1929. Kringlen studied all "nonorganic" psychoses, including schizophrenia. Unlike the Finnish project, he used males and females and studied opposite-sex as well as same-sex fraternal twins in extensive fieldwork, reporting no difference in these fraternal concordance rates with respect to schizophrenia.

Kringlen found an identical-twin concordance rate for schizophrenia of 60 percent for the most severely ill part of his sample, dropping to 25 percent for the most benign, uncorrected for age. His study also looked at how specific concordance could be for a particular psychosis. Of 21 psychotically concordant pairs, 20 cotwins had the same form of illness: schizophrenic, reactive, manic-depressive, or schizophreniform psychosis—strong evidence against the "continuum of psychosis" notion espoused by Karl Menninger and the British psychiatrist, T. J. Crow. With continued follow-up, the schizophreniform group has been merged with the schizophrenic group of twins as one category and reported in Table 8.

In Denmark, by matching the National Psychiatric Register with the National Twin Register, the psychiatrist Margit Fischer, a student of Munich-trained Erik Strömgren, and geneticists Bent Harvald and Mogens Hauge constructed a valuable national psychiatric twin register. From a starting sample of nearly 7,000 same-sex pairs born between 1870 and 1920, they found 395 psychiatrically hospitalized pairs of whom 21 identical and 41 fraternal twin pairs were schizophrenic. Fischer followed the subjects for more than 24 years after the first twin in a pair became ill; such a design eliminates the need for age correction of the concordance rates.

The Danish study's concordance rate in fraternal twins, 27 percent, is very high, but it can be accounted for partly by the severity of illness in the index twins and by the possible misclassification of a few concordant MZ pairs as DZ pairs. One undesirable result of the long study period was that 35 percent of the original subjects were dead when final results were collected, so there was no chance to personally interview a substantial portion of the original group. On the other hand, this long period reproduced an important finding. Most concordant identical twins became ill at about the same time, but Fischer still found two concordant pairs with 17 and 29 years of difference between onsets — impressive clinical evidence for some kind of environmental triggers or releasers.

Fischer's study attended to all personality abnormalities in cotwins ranging from neurotic to nervous to odd. When all of these possible spectrum abnormalities, from schizophrenic to odd, were considered to be positive evidence of being affected with schizophrenia, concordance jumped to 64 percent for MZ twins and 41 percent for DZ twins. There were no conspicuous sex differences in concordance, and this study found no correlation between concordance and severity. Fischer suggested that the relationships between severity and concordance in the twin literature stemmed from environmental rather than genetic factors. We suggest, in addition, that the narrowness of Fischer's diagnostic standard may have limited the range of severity and thus the correlation. The Danish researcher looked carefully for environmental differences that could account for one twin in an MZ pair being schizophrenic and the other being well; she concluded that such factors were idiosyncratic where they could be observed, such as head injury or the reaction to menopause. The case histories provide challenges to idealogues of either hereditarian or environmentalist persuasions.

A nation's twins constitute a marvelous national treasure and resource for biomedical research, as we see from the national twin registers kept over many years in the Scandinavian countries. It is not generally appreciated that the United States has a special twin register of its own, and, even though it is not complete and contains no women, it contains a wealth of information that can advance our knowledge about the human condition. All male twins born in 39 of our 50 states between 1917 and 1929 (54,000 multiple births) were screened through the Master Index of the Veterans Administration (VA) for all twin pairs where *both* members had served in the Armed Forces; 15,924 pairs of twins were found. (A Vietnam-era twin register has been initiated at the suggestion of the author.) When the available records on these twins were screened for chart diagnoses of a broad, DSM-II-like construct (synonymous with the old dementia praecox) of schizophrenia, 164 pairs of identical twins and 277 pairs of fraternal twins were identified. All data analyzed for this important study were from VA charts; none of the investigators saw or interviewed any of the twins.

It is a screened sample of "schizophrenics," screened for both physical and mental health in both members of a pair, and the exact consequences for concordance rates are difficult to evaluate. The results of the study are not averaged in with the studies in Table 8, all of which involved extensive and intensive contact between the researchers and the twins. We know from public records that 1.7 million men were rejected at the time of induction during World War II for "emotional or mental" reasons, and a further half million were separated from the services for these reasons after induction to active duty.

The psychiatrically disordered twins were initially studied by William Pollin and colleagues at the National Institute of Mental Health (NIMH) in 1969, but that report has now been superseded. In 1981, Kenneth Kendler and Dennis Robinette reported on an enlarged sample after a further 16-year follow-up of the records kept by the Veterans Administration. By the time of the follow-up, all the twins would have passed through the risk period for developing schizophrenia. The probandwise rates for the broad definition of schizophrenia were 31 percent in the MZ pairs and 6 percent in DZ pairs. By broadening the definition of concordance in cotwins to include *any* kind of psychiatric diagnosis (mostly personality disorders), the rates grow to 53 percent for the MZ pairs and 24 percent for the DZ pairs. We will never know the "true" rates for "true" schizophrenia and should not expect them from an essentially epidemiological approach to psychiatric genetics. Only a third of the diagnoses of schizophrenia would meet criteria for DSM-III schizophrenia, according to Kendler, but another third would most likely meet criteria for a reasonable definition of the schizophrenia spectrum.

The twin study we lean upon most heavily for our insights reports on the British data collected by the author of this book and his coinvestigator, the late James Shields of the United Kingdom. In 1948, Eliot Slater, whose own twin study is described among the classical ones above, initiated a new twin register designed to avoid sampling biases that is maintained at the Maudsley-Bethlem Hospital and Institute of Psychiatry in London. Sixteen years later, Gottesman and Shields collected follow-up information from fieldwork and hospital interviews to select 62 schizophrenic subjects from 57 pairs of twins from a total register of 479 disordered twins — the total yield from about 45,000 consecutive psychiatric admissions. Unlike most schizophrenic samples, this one had a higher representation of cases with a good prognosis, because subjects were selected from consecutive outpatient admissions as well as from consecutive inpatient admissions. As it turned out, however, all the outpatient schizophrenic twins had been hospitalized by 1965, even if only for a few weeks. The Maudsley-Bethlem study was designed to answer most of the criticisms of the earlier twin studies in regard to genetic biases, sampling biases, and diagnostic uncertainty.

In Chapter 2, we described how this sample was refined by using the blindfolded diagnoses of a panel of six judges, rather than chart diagnoses or our own evaluations. That removed two pairs of identical twins from the study. The concordance rates resulting from the six-judge consensus were 58 percent for MZ pairs and 12 percent for DZ pairs at the level of schizophrenia-like "functional" psychoses. These rates are not age corrected because the discordant twins were followed initially for 3 to 27 years. Age correcting would have raised the rates artificially because the healthy cotwins had, in effect, already used up their risk for a future schizophrenia. Even though a number of these original subjects were still within the risk period for developing schizophrenia, a further check of national health insurance records (with the aid of Dr. Adrianne Reveley) after 20 more years (1985) found only one new schizophrenic in the total sample, and she was a cotwin from a DZ pair, raising the rate 3 percent to that reported in Table 8.

Challenges to Received Wisdom

If the results from earlier twin studies could be attributed largely to *clinical* diagnoses based on unstructured interviews or to skimpy military records made by nonspychiatrically trained doctors, we might expect, if we were skeptical of genetic interpretations, that the strong evidence in favor of the role of genetic factors would become weaker when objective stipulations are imposed. We recycled the Maudsley-Bethlem schizophrenic twin sample and applied current objective criteria sets blindly to the original case histories, including verbatim transcripts of semistructured interviews (see McGuffin et al. 1984, and Farmer et al., 1987). The raw sample size yields of schizophrenics thus diagnosed ranged from 3 [sic] to 54; clinical diagnostic yields had ranged from 34 to 81, with the six-judge consensus identifying 69 cases of schizophrenia. The different concordance rates generated by objective criteria applied blindly with regard to zygosity now become of interest to extract genetic meaning.

Research Diagnostic Criteria (RDC), the objective rules for making a psychiatric diagnosis that served as the model for DSM-III, preserved the MZ sample size but excluded one-third of the DZs, yet still resulted in concordance rates comparable to the older but clinically based rates: RDC rates in MZ pairs of 46 percent and in DZ pairs of 9 percent. Applying the DSM-III criteria led to virtually identical concordance rates of 48 percent in MZ twin pairs and 10 percent in DZ pairs, with reductions in sample size similar to those for the RDC criteria. Requiring one or more Schneiderian FRS to be specifically described in (rather than inferred from) the case history reduced the sample to a trivial 17 twin probands and produced, paradoxically, higher concordance rates for DZ pairs than for MZ pairs. Based only on these results for a narrow definition of schizophrenia, genetic factors would be seen as *not* important in the ctiology of schizophrenia. Extending the search for the most genetic definition of schizophrenia — a definition that would maximize the ratio obtained by dividing the MZ by the DZ concordance rate for a particular definition — and with Dr. Crow's guidance, we were able to divide the twins into his pure Type I, characterized by positive symptoms (delusions, hallucinations) and favorable response to medication; his pure Type II, characterized by negative symptoms (apathy, withdrawal, loss of pleasure responses), morphological brain changes, and treatment resistance; and a mixed type. Only three twins were Type II, thus casting doubt on the usefulness of that concept. Probandwise concordance rates for Type I were 53 percent (8/15) in MZ pairs and 19 percent (4/21) in DZ pairs; for the mixed type, they were 64 percent (7/11) and 0 percent (0/11). Our initial attempt to search for genetic meaning in Crow's typology yields encouraging results — an MZ/DZ ratio that is very high for the mixed type — as well as discouraging results — almost no cases diagnosed as Type II — but these data could be interpreted as reflecting a continuum of severity within a broader category of schizophrenia.

Severity as an Indicator of Genetic Loading

Gottesman and Shields also explored the relationship between severity of illness and concordance in several different ways. When the test of severity was either a twin's inability to work or hospitalization within the six months preceding the follow-up interview, the concordance rate for severely affected identical twins was 75 percent and the rate for mildly affected identical twins was only 17 percent. When subtype diagnosis of nuclear versus nonnuclear schizophrenia was the severity criterion, the severe MZ rate was 91 percent and the mild MZ rate was 33 percent. If the personality of the subject before the illness (the presence of a pre-morbid schizoid personality) was the criterion of severity, the results were equally striking: 82 percent for MZs with schizoid personalities, 25 percent for MZs without.

Reprise: Old and New Twin Studies

We believe that the pooled probandwise concordances for schizophrenia in Table 8 give the most accurate picture of risk among identical and fraternal twins for industrialized societies in the last half of the twentieth century. Given the perspective discussed above about design differences between the classical and the new studies, we see the entire array of twin findings as fulfilling a basic requirement of hard science — replication. Undeniably, when twins are reared together, the identical twin of a schizophrenic is much more likely to be schizophrenic than is the fraternal twin of a schizophrenic. No one or two concordance rates provide the last word from twin studies, but the findings are consistent and very orderly. Excluding the VA chart-based study, the pooled data in Table 8 result in an identical twin concordance rate of 48 percent and a fraternal concordance rate of 17 percent. These results indicate that shared genes certainly have something to do with the source of schizophrenia. However, the fact that only half of the identical twin pairs are concordant for recognizable schizophrenia tells us that something important beyond genetic factors is required to complete the solution to the puzzle of causation.

There is reasonably good agreement among the recent twin studies of schizophrenia. The pictures are similar partly as a consequence of using the probandwise method of calculation and partly as a result of thorough sample investigation and longer follow-up. But they are similar also because modern investigators are agreeing more on diagnostic criteria and taking the middle of the road more often. We favor a standard that accepts as concordant, cotwins who are schizophrenic or probably schizophrenic, virtually all of whom have been hospitalized for a functional (i.e., not obviously organic) psychosis, one that is not likely to be an affective psychosis. For the second phase of data analysis, we would add the cotwins with genetically relevant schizophrenia spectrum disorders, but currently there is no sure way to identify such conditions. Developments in brain imaging, in cognitive neuroscience, and in molecular genetics should soon remove many of the ambiguities of clinical diagnoses that rely only on behavior.

Understanding Discordance in Identical Twins

In nearly every twin study of schizophrenia, half the identical twin pairs contain a cotwin who is not overtly psychotic. These discordant pairs have been intensely scrutinized because they offer such seemingly fertile ground for finding out what environmental factors influence schizophrenia. Two persons of exactly the same genetic makeup reared in the same household have radically different mental health outcome statuses. Surely, somewhere in the experiences of these individuals lie revelations about specific stressors, viruses, ecological encounters, or predisposing factors that contribute to the illness or, on the other hand, factors that prevent an individual from succumbing to schizophrenia.

Loren Mosher, William Pollin, and James Stabenau, all then of the NIMH (1971), conducted a nationwide study of 15 MZ pairs selected to be discordant for schizophrenia. They examined birth weight, identification with parents, submissiveness and dominance within pairs, and several other neurological and psychosocial variables. Such information, along with data from other studies, doesn't support the idea that birthweight differences or other birth-related difficulties are *specifically* connected with schizophrenia, but such factors have been implicated as risk-increasing ingredients in the multifactorial stew. Within-pair submissiveness does correlate in a number of the twin studies with which of two genetically identical individuals becomes schizophrenic, yet is submissiveness a cause or an early-appearing effect? We don't know.

Despite high hopes, the study of discordant identical twins hadn't yet led to any big payoff, but new techniques in biology suggest there is still gold in those hills. The problem is simply more difficult than had been imagined. The environments of individuals sharing one household differ in small ways that are difficult to specify. Besides, there aren't many of these special twin pairs who are available and willing to be studied in the intensive manner required. Our questions about them are subject to recollections that easily could be distorted by what has happened since one became ill. Perhaps the things that we are looking for occurred long ago, even prenatally; are very nonspecific; or vary greatly. As if these problems aren't enough to confound us, there are plenty of others. Even the decision about which twins are discordant and which concordant is open to question. The decision is made by drawing an artificial dividing line on a scale of behavior that ranges along a continuum from "normal" to "schizophrenic." Individuals who are very close together but on opposite sides of that arbitrary line are called normal *or* schizophrenic. The problem can be illustrated with a few case histories from our British sample. These are all twins who were judged to be discordant (six-person consensus) for schizophrenia.

In one pair, the schizophrenic twin probably had an organic psychosis that resembles schizophrenia, a so-called symptomatic psychosis (see Chapter 2). He was a Japanese prisoner of war and may have sustained nutritional brain damage during periods of starvation in the POW camp. He is deaf, and his electroencephalogram — EEG (brain-wave tracing) — suggests a brain lesion. His twin is married and has a stable job history; personality testing was within normal limits, but he was very defensive about admitting anything unfavorable about himself; further, he has not suffered the insults of war, malnutrition, and deafness. He refuses to be interviewed, however, and may have some kind of spectrum condition — but we can't prove it. If the proband really has a phenocopy of schizophrenia (an imitation of the real thing), the cotwin can provide no information relevant to the causes of "true" schizophrenia. Alternatively, the proband may be like those schizophrenics studied by Schulz (pp. 94–96) whose psychoses were precipitated by a somatic insult to the brain and whose relatives had very low rates of schizophrenia, but nonetheless higher than the general population risk of 1 percent.

In a second pair, the schizophrenic twin has been hospitalized only once for six weeks. In this case, both twins are fascinated by the occult. The schizophrenic twin joined a messianic cult at age 20, and two years later left his second wife for special indoctrination. During this time, he underwent what he called "psychic realignment" and was trained in telepathy. He reported that we was "radiating his thoughts" and felt that other cult members knew the details of his sex life. In our conversation at follow-up four years later, he reported that we was working at two jobs to make ends meet, was back together with his second wife and child, and was no longer on medication, but felt depersonalized and had auditory hallucinations. He said he was a "radiating telepath," but had enough insight to conceal this information from his coworkers. His brother was, at first follow-up, quite strange. At 20, he was married, had children, and held a responsible position. He told me that he, like his brother, was a member of the same occult society and had left his job to work at its national headquarters. He compared himself to Paul of Tarsus. He considers himself too careful to become insane, more self-reliant than his twin, and physically and mentally "stronger." On personality tests, both twins show paranoid and schizoid traits that we associate with schizophrenia when it is in remission. Blindfolded judges considered the second twin to be within the "schizophrenia spectrum" when reviewed at age 26, although he had never been treated or hospitalized by age 47. This pair represents an instance of the arbitrariness of the threshold for "normal" and of the frustration of hunting for releasers or protective factors.

One final example includes a schizophrenic twin who had been hospitalized three times, first at age 16 when he was diagnosed as undergoing an adolescent depression. He was once, however, diagnosed as having a schizoaffective psychosis. Electroconvulsive therapy (ECT) cleared his depression, but his thinking disorder remained; numerous episodes of schizophrenia have ensued. Drug therapy with antipsychotic medication has improved his condition enough for discharge to outpatient status. His brother, who has had no contact with the psychiatric world, hospitalized himself twice for physical complaints ("stomach ulcers") with no detectable organic basis. He has a poor work history and was the poorer student in school. The nonschizophrenic twin has been more exposed to their unstable mother (she fits the mythical schizophrenogenic image) and harsh stepfather, whereas the ill twin has lived mostly with a kindly maternal relative. Although the nonschizophrenic twin has personality-test traits that resemble a clinical case of schizophrenia, he does not qualify on behavioral grounds for the diagnosis of schizophrenia, but belongs somewhere in the spectrum. As with the second pair discussed above, we may have evidence here of the partial expression of the genotypic potentiality for developing schizophrenia, a view in agreement with one that assigns a role to regulatory factors, either other genes or random environmental conditions.

Revitalizing Research on Discordant Twin Pairs

Much mileage remains, however, in the strategy of carefully scrutinizing the biology and the psychology of identical twins who have remained discordant for five or more years. With the advent of the new techniques for brain imaging — such as MRI, rCBF, PET, and fine-grain analyses of the EEG (BEAM) — and the ability to store DNA itself for analysis with future gene probes (restriction fragment length polymorphisms — RFLPs), E. Fuller Torrey, the eminent psychiatrist and patient advocate, and Gottesman have initiated a new nationwide study of discordant schizophrenic identical twins with colleagues at Saint Elizabeths Hospital (NIMH). The twins, more than two dozen pairs, are brought to Washington, D.C., with their parents to undergo a week-long series of evaluations with the new techniques and the old (e.g., careful life-span histories of significant events and personality assessments). We are hoping to see what actually differs between the brain of a schizophrenic and that of his or her genetically matched twin who is currently functioning well. At the same time, control information is being obtained from pairs of MZ twins who are concordant for schizophrenia and from pairs of MZs who are both normal. The study will continue into the 1990s. (If you know of discordant pairs of identical twins, please let us know by writing to Twin Unit, WAW Saint Elizabeths Hospital, Washington, D.C. 20032.)

Identical Twins Reared Apart

The differences in the environmental experiences of identical twins discordant for schizophrenia seem insufficient to account for the discordance. Not many low-birth-weight babies, nor children tied to the apron strings of constricting and inconsistent mothers, nor submissive twins who have difficulties in fending for themselves because of their close relationship, actually develop schizophrenia. Such factors, however, may be important in determining illness and health among identical twins who have a predisposition to schizophrenia. For some scientific problems, the occurrence of identical twins reared apart from birth seems to give us an ideal "natural experiment" to resolve some of the conflicts over the relative importance of nature and nurture. For schizophrenia, however, these cases are very rare and the circumstances that create them are unusual, to say the least.

Fourteen pairs of identical twins reared apart, one of whom has been diagnosed as schizophrenic, have been described in the scientific literature, but the sample is too small for generalization (Gottesman and Shields, 1982). Of these, however, 64 percent are concordant, a value somewhat higher than the concordance rate of 48 percent for identical twins reared together. In evaluating these data, we must ask ourselves how the circumstance of one twin or both being given up to be reared in different families is likely to have come about. We probably want to look at this assembly of pairs as little more than fascinating curiosities. It's the kind of thing about which we should say, Lo and Behold! — and then get on to other matters.

One of the few examples of concordance for schizophrenia in twins reared apart was found in Gottesman's and Shield's London study. Herbert and Nick (not their real names), the children of a 19-year-old half-Chinese woman impregnated during a casual encounter with a British repairman whom she never saw again, were separated at birth in 1934. They stayed in two different series of foster homes as infants, but were reunited at age 5 during the evacuation of children from London to avoid the German air raids; after less than a year they were separated again, and Herbert only was reared by his Chinese maternal grandmother. Because the grandmother did not want the added burden of Nick, she gave him away to a 41-year-old childless woman whom she knew (the foster mother of the twins' aunt), whom we will call Mrs. M.

This narrative becomes even more like a soap opera. Mrs. M's husband was not consulted, but acquiesced in her decision to take Nick, who was then 7-½ years old. A year later, at 42, Mrs. M. became pregnant. Herbert continued to live in the Limehouse area of London with his grandmother and her second Chinese husband, who spoke little English. Neither paid much attention to Herbert. In contrast, Nick lived comfortably in suburban Anglo-Saxon London with caring, warm adoptive parents and a compatible younger stepsibling.

Herbert had been the smaller and more delicate of the twins. After his mother had been in labor for two days, only Herbert had a forceps delivery. He had probable head concussions at ages 10 and 12. Herbert was an enuretic (a bed wetter) until he was 12, as well as a fire setter and dog torturer because, he said, "the devil told me to." After thievery at age 15, he was committed to a school for delinquents for one year. He was drafted into the army and served in a unit for "dull soldiers"; he was found to be illiterate, with an IQ of 87. Nevertheless, he served honorably, never meeting his twin who had also been drafted into a similar unit.

Nick did poorly in a private school and from age 10 to 14 was put into classes for slow learners; he wet the bed until age 14; he was brought to police attention for setting fires at ages 9 and 10 and was twice convicted of theft at age 17 and sentenced to a hostel on probation. He too was drafted into the army and found to be illiterate, with an IQ of 75; he also served honorably, doing menial tasks. After discharge, he worked briefly but quit because "fellows were spying on me." Despite their illiteracy and low IQs, both could carry on quite intelligent conversations with me. Neither twin was ever interested in girls.

Both Herbert and Nick held various unskilled jobs as delivery boys until age 22. Herbert then began to behave oddly, staring silently into space, sitting in awkward positions for long periods, neglecting himself, grimacing and laughing to himself; he interpreted passing automobiles as the sound of enemy aircraft. Nick's family heard that Herbert was not well and thought a visit might help; the twins had rarely seen each other since age 7½, but were aware of each other. The visit took place on December 22, at which time Nick learned that their biological mother had seen Herbert and their grandmother briefly in October while visiting the country from her home in the United States.

On January 8, Herbert was admitted to our hospital and came onto the twin register: "You feel people are deceiving you. . . . I'd be reading people's thoughts when I concentrate. . . . Some people talk backwards and some people you have to get along on top of their talk. [Later] I'm sure an 'interdiscrete society' could help you. Communist aggression mixed with racial intolerance . . ." Herbert was committed to long-term care in a mental hospital and was still there after more than 28 years.

Unknown to us at the time, Nick was admitted to a different hospital on January 5 after running across a plowed field with his arms outstretched as if in prayer. The night of his visit to Herbert, he was found crying and the next day seemed lost in thought and was making clicking sounds with his tongue. After New Year's Day, he amazed his adoptive father with unintelligible talk; he felt that he had special powers but that they left him when a cigarette pack was thrown away; he smashed a porcelain dog — "The devil was there and it was either him or me"; he saw a mass of flames and heard voices singing "Hark, the Herald Angels Sing." He was admitted to a mental hospital the next day in a confused and agitated state. Like his twin, he had been virtually continuously in hospital for over 27 years, with no contact with his twin, when he was killed in an automobile accident on the hospital grounds. Both did better on different phenothiazines than without them, but neither had ever improved enough for discharge. During follow-up in 1963, I interviewed their maternal aunt who had been admitted for paranoid schizophrenia at age 43, five years after either of the twins.

Certainly not all the twins reared apart are Herberts and Nicks, but their circumstances are also unusual in their own ways. So, though fate seems to offer wonderful research possibilities, the cases are often too odd to permit generalizations. Although reared apart in quite different circumstances, they both became schizophrenic at age 22; was it stress, was it their biogenetic blueprint, or was it some kind of interaction? No doubt partisans can find the facts in this particular story to fit their preferred, if opposing, theories.

The Offspring of Schizophrenic Twins

Studying the adult offspring born to concordant and discordant MZ and DZ twins produces unique evidence for testing the idea that the genotype for schizophrenia can be completely invisible or unexpressed in a "carrier" and yet be transmitted to the next generation. If the risk of schizophrenia to the offspring of schizophrenic MZ twins is like the risk to the offspring of ordinary schizophrenics, about 13 percent, and if the risk to the offspring of the clinically healthy MZ cotwins is close to the population rate of 1 percent, it would provide very powerful evidence *for* the role of the environment (the rearing factors associated with a schizophrenic parent) in causing schizophrenia. Alternatively, it would raise further questions about nongenetic phenocopies of schizophrenia in the ill MZ twin. If, on the other hand, the risk to the children of the healthy MZ cotwins matches that in the children of the sick MZ twins, it *undermines* the significant causal role for shared rearing factors such as social class and for *the unshared factor of a schizophrenic parent*.

The offspring-of-twins strategy was introduced to psychiatric genetics in 1971 by Margit Fischer for her Danish schizophrenic twin sample described earlier in this chapter. The sample was quite young at that time and did not contain as much information as would be revealed by a follow-up study when the offspring of the twins were well into or through the risk period for developing overt schizophrenia. Another advantage of a follow-up would be to confirm the healthy status of the discordant cotwins who might have had a late onset of schizophrenia compared to their ill twin. Aksel Bertelsen, the Danish psychiatrist who had conducted the definitive twin study of manic-depressive disorders, and Gottesman completed (1989) a follow-up of the Fischer sample of twins and their offspring, adding 18 years of further observation, including that of the fraternal twins. This kind of research could have been conducted only in Scandinavia, where we had to access to the unique system of national records covering 120 years that we needed to evaluate the psychological statuses of the twins and their children embedded in a population of five and a half million citizens.

Table 9 shows the number of twins with offspring from the starting sample of Fischer's 21 MZ and 41 DZ pairs (124 individual twins) described earlier. The healthy twins had passed through the entire risk

TABLE 9

Gottesman-Bertelsen follow-up (1985) of Fischer (1966–1967): number
of reproductive twins, by zygosity and diagnosis

	Schizophrenic Twins		Normal Twins	
	Reproductive	Total	Reproductive	Total
MZ	12	31	7	11
DZ	11	49	21	33
Total	23	80	28	44
Reproductive rate	29%		64%	

After Gottesman and Bertelsen (1989).

period, so we are quite confident that they never became schizophrenic.
Note that less than a third of the schizophrenics had children, compared
to some two-thirds of their well cotwins. The twins had a total of 150
children, of whom 14 died young but of whom 115 were over age 35 at
follow-up.

Table 10 shows the remarkable findings from this very special combination of twin and family strategies. The risks of developing schizophrenia for the offspring of the two groups of MZ twins are quite high and are
indistinguishable from each other. The risk for the offspring of schizophrenic DZ twins is virtually the same as for the two groups of MZ twins.
The risk for the children of healthy DZ cotwins, actually the nieces and
nephews of schizophrenics, is close to that reported in the literature (see
Figure 10) for such relatives, only 2.1 percent.

The data we have shown from this unusually well-followed-up sample
of Danish schizophrenic twins and their offspring support a strong role
for genetic factors in the etiology of schizophrenia. No support is found
for the suggestion that rearing by a schizophrenic parent is necessary or
sufficient to produce schizophrenia in offspring. Our data do not support
the hypothesis that nongenetic factors such as viruses, cerebral injuries,
or toxins frequently lead to monozygotic discordance, because the offspring of the phenotypically normal MZ cotwins of schizophrenics had
the same risk of developing schizophrenia as if they were the children of
schizophrenics. It would appear that the genotype for schizophrenia can
often remain completely unexpressed and silent to clinical detection, be
transmitted, and then be expressed in the next generation when the sons
and daughters encounter a different constellation of relevant stressors.

TABLE 10

Schizophrenia and schizophrenia-like psychosis in offspring
of schizophrenic twins

		Monozygotic Sample	
		Offspring	
Index parents	Number	Schizophrenia and schizophrenia-like	Morbid risk (%)
Schizophrenic twins N = 11	47	6	16.8
"Normal" cotwins N = 6	24	4	17.4
		Dizygotic Sample	
		Offspring	
Index parents	Number	Schizophrenia and schizophrenia-like	Morbid risk (%)
Schizophrenic twins N = 10	27	4	17.4
"Normal" cotwins N = 20	52	1	2.1

After Gottesman and Bertelsen (1989).

Identical Quadruplets Concordant for Schizophrenia

No account of the role and value of twin studies of schizophrenia could be complete without a mention of the amazing story of the Genain quadruplets. They can serve as a microcosm of all schizophrenia research. David Rosenthal (1963) was seduced and initiated into the complexities of the debate over the role of nature and nurture in schizophrenia when he was presented with the opportunity to coordinate the intensive study of identical quadruplets, each schizophrenic, brought to the Clinical Center of the NIMH for intensive scrutiny over a period of years with the consent of their parents. The clone of four female schizophrenics is ideal for generating hypotheses, since they differed in symptoms, onset, course, and outcome despite their identical genetic predispositions for developing a schizophrenic phenotype. All were grossly exposed to the same pathological family environment and all became ill; but a combination of circumstances, among which the different patterns of interpersonal relationships within the family were prominent, led to pictures ranging from mild with recovery to very severe with a deteriorating course.

Some 25 years after the initial study, the women were recalled and restudied with an armamentarium of high-technology devices in the hope of supplementing the clinical information already in their record. PET scans, tests for neurochemical variation, MRI images of the brain, and neuropsychological tests were applied. The major lessons to emerge (see the papers by Buchsbaum, DeLisi, and Mirsky in 1984) concern the clinical variability possible for just one example of a schizophrenic genotype cloned four times. Such variability highlights the important roles played by psychosocial and nongenetic biological factors, as well as the timing of the genetic regulatory factors when there is no variability in the genes themselves (at least at the time of the initial four-way splitting of the fertilized ovum). PET findings suggested that frontal brain hypofunction (diminished activity) was not familial, but may be a "state marker" reflecting their clinical condition at the time rather than an indicator of a more durable genetic trait marker. Ventricle-to-brain ratios (VBR, an indicator of "holes in the brain") appeared to be familial but, contrary to expectation from the literature (see Chapter 12), the MRI scans from the quads did not show ventricular enlargement or any atrophy of the brain cortex. One hopeful finding was the lack of deterioration after 25 years in these pharmacologically treated, cared-for, chronic schizophrenics (three of them) or schizoaffective (the remaining one) quadruplets. The exciting strategy of multiple biological and psychological assessments by a team of scientists has much to recommend it and is the protocol being followed by Torrey and his colleagues at Saint Elizabeths Hospital with a new, nationwide sample of identical twins discordant and concordant for schizophrenia.

Conclusion

The facts about the risks of schizophrenia obtained from family studies of schizophrenics — their parents, siblings, children, and more distant relatives — all suggest that schizophrenia is familial; that is, it is much more common among relatives than among members of the general population. One line of evidence that the familiality is the result of sharing genes, rather than experiences, is the classical genetic strategy involving the comparison of concordance rates in identical and fraternal twins of schizophrenics, together with variations in that method, such as the microanalysis of discordant identical twins, twins reared apart, and a set of quadruplets concordant for variations of schizophrenia. The observed rates in twins and other relatives suggest that genetic factors are important, though not adequate to explain all the observations. Still unclear is the role played by possible shared experiences within the family, for which the study of children adopted away from their schizophrenic parents is required.

The case history that follows returns us to clinical reality and to the raw material for genetic analysis. The problems faced by the family members show how much farther we must travel on the research road before we can make recommendations about treatment and prevention.

Three Generations of Schizophrenia

A social worker whose uncle, brother, and daughter are all schizophrenic shares her anguish and perspective.[1]

Ibsen treated it tragically in *Ghosts* and Gilbert and Sullivan humorously in *Ruddigore*. I am referring to the theme of a family curse in the form of insanity passed on from one generation to another. I disliked *Ghosts* because of its morbid outlook and enjoyed *Ruddigore* because of its lighthearted foolishness. In neither case did I think that the idea of inherited mental illness was realistic. Years later, I recognize that during all that time my family was enacting its own tragedy affecting several generations. Although I never chose it, a small part was thrust upon me as a child and I was inexorably drawn into a larger role as the tragedy progressed. I am referring to schizophrenia.

Early in my life I became aware of the visits my father made to his brother. The trip was an exhausting 3-hour train and bus ride to a sad and mysterious place. So sad that my mother, whom I knew to be kind, was unwilling to go. She did, however, spend the previous day cooking the

---

[1]L. Fuchs, *Schizophrenia Bulletin*, 1986, 12: 744–747.

meal my father was to share with my uncle. Early in the morning on the day of the trip, I watched my mother and father pack additional delicacies, mended underwear, and miscellaneous items like magazines and drawing supplies chosen to entice my uncle into activity. The day after was given to a review. "What did the doctor say?" made sense to me. "How did Aaron behave?" was disquieting as it implied misbehavior and childishness in an adult. As I grew older, I observed my father become more exhausted from his trips and more and more disappointed in his brother's failure to improve. The saddest visits were when my father had to report that my uncle's behavior led to being moved to a "bad ward" [probably a back ward for chronic patients with little hope for improvement]. At the end of the 1940s, my parents deliberated about a proposed lobotomy. They finally agreed, hoping for the best. The best never happened. My uncle died in Pilgrim State Hospital in his eighties. He had been there for 60 years.

I did not see my uncle until I was an adult, but his self-portrait hung in our living room. His handsome, brooding presence made me feel a loss, for though I had other uncles, no other was an artist or so young.

And what caused this sad illness? Without knowing the name of my uncle's disorder, we all seemed to have accepted psychoanalytic thinking. My grandmother, my uncle's mother, was a dignified but cold woman so that the belief that schizophrenia was due to maternal rejection grew on fertile ground.

My uncle's condition had a name and came on suddenly. More tragic was my brother's, which had no name and grew insidiously. A bright, handsome child, 5 years my senior, he was loved by my parents and was a delight to his teachers. He was not, however, interested in children. As he grew older, my parents became increasingly anxious about his preference for adults. I shared their anxiety and, 40 years later, I remember our enthusiasm when he spoke of two boys whom he might invite home. Although they never came, I remember their names still. So, in subtle ways, the natural order was reversed, and I was asked to understand and make allowances for my older brother.

Lacking his own friends, he bossed and teased me. Although he played with me rarely, he resented my play with other children and disrupted our games or was embarrassingly disagreeable when I brought new friends home. Hardest for me were his sudden changes of mood. I risked his anger when I did not follow his direction and my own self-respect when I bowed to pressure. These choices were all the more painful because I loved him. As he reached adolescence, he was also becoming more critical and angry toward my parents. The three of us became adept at avoiding confrontations. Things reached a climax when he was 16. He started attending a local college and began having trouble studying. Now I know one reason was the schizophrenic "loss of thought." My parents arranged for a psychiatrist to see him, but after one session he refused to return.

Despite these problems, my brother must have made heroic efforts. Not only was he graduated from college, but he even served in the army for a few years. He never saw active duty and was taken under the wing of his superior officers in an Intelligence Unit. They appreciated his dependability and strict and religious outlook. Despite warm and lengthy letters, he was again difficult to live with on his return. He completely disowned me because he disapproved of my engagement and did not speak to me until my mother's terminal illness 15 years later. During this time, he structured a life with which he could cope. He never married, became intensely religious, and spent much time and energy in helping our aging parents. He worked as an elementary school teacher in an inner city school and won some acclaim as a strict but fair teacher to children who needed rules and structure. He finally found friends within a strict religious community, and this was some comfort to my parents who had by then resigned themselves to his not marrying. They derived some satisfaction in knowing that he was involved in his work and at long last had friends. I am detailing my brother's long-range adjustment because I believe attention must be paid to other "mild schizophrenics" who are unhappy and cause pain to their loved ones but still manage a life outside treatment and without an official diagnosis. Eventually my brother diagnosed himself. His anger abated as his delusional thinking loosened its hold. He no longer sees me or my husband as evil people. He divulged his self-diagnosis when my daughter became psychotic.

My brother wanted to be helpful and told me that he noticed his own thinking became less difficult when he was taking medication for another illness. He shared with me how he talked back to his voice when it caused trouble. "If you can't say something wise, shutup!" Before this revelation, I had not formed a definite opinion of his problem. I considered schizophrenia as a diagnosis when he was unreasonably accusatory and punishing toward me, but leaned toward obsessive-compulsive disorder when his accusations had some basis in reality. At any rate, in the light of his own experience, he encouraged my husband and me to continue our sympathetic support of our delusional daughter, and stressed the need to be hopeful and encouraging. So, family commitment has survived years of pain, madness, and confusion and the years to come hold some promise of affording simple family contacts and pleasures.

If diagnosis eluded me for so long, so did causality. Though sympathetic toward my parents, I was also influenced by the thinking of the time and in my heart I blamed them. Now I realize that living with a mentally disturbed son had made these naturally gentle and cautious people increasingly anxious. At the time, it was to easy to credit neighbors' and relatives' comments that my brother's problems were caused by my parents' insecurity and their foolish capitulation to his angry outbursts. Today, some family therapists lean dangerously close to this attitude with their theory of the "identified patient" who acts out the pathology of the rest of the family. True in particular cases, perhaps, this theory unfeelingly ignores the inevitable changes that occur in families who live with a mentally ill member. At any rate, when I was young, I resolved that as a parent I would neither neglect one child nor indulge the other, and I was determined to live life with more joy. Good resolutions —but no proof against biological or genetic catastrophe.

Considering my early life, it is clear why I would be drawn to working with troubled people and why I would also look forward to having a happy, healthy family. Things went well, and my husband and I were confident about our children's futures in their early years. I knew, of course, that there was no perfect child and that there was a certain amount of pain and struggle in every life. For this reason, when my second child, Susan, turned out to be an unusually quiet baby and later a shy toddler, I valued these qualities as part of her unique nature. If people were less attentive to her than to her more outgoing older brother, I wondered at them and cherished her all the more. Looking back, I can see that very early I tried to fill the gap between her and the rest of the world by becoming more attentive, playful, and protective of her. So, slowly and quite naturally, I accustomed myself to her ways and to feeling responsible for her happiness. Despite her early reserve, by 7 or 8, she had become an energetic and interested child. She excelled in school, was an avid reader, and did well in many athletic activities. With all this, she still did not find making friends easy. She was shy, easily hurt, yet critical and argumentative. We thought she would eventually find a special friend or two, and we minimized her legalistic quarrels by thinking she had strong opinions and was independent. Jokingly, we said she was surely cut out to be a lawyer.

By high school Susan was more outgoing. She attracted potential friends because of her intelligence and her dynamic interest in ideas and activities. She was still easily hurt, however, and her relationships were uneven and stormy. It was a period of confusion and stress for me as well. Although I was the one she came to in tears for comfort, I was becoming aware that there was a lack of warmth and reciprocity. We no longer had the good times together that we had earlier. There was no balance, and living with my daughter became emotionally draining. Still, I accepted the common wisdom that Susan was merely going through adolescence and that I was an overly concerned mother. When Susan reached age 15, we finally sought professional help because of the appearance of a new symptom as she began to have trouble completing written assignments. Once or twice she said she thought she was going crazy, because she couldn't think. That was outside the realm of my imagination, and I assured her that most adolescents have these fears. It was apparently outside the range of possibilities for the psychiatrist as he indicated Susan had some problems in self-esteem and suggested my distancing myself more. I followed this advice as well as scrupulously keeping out of the treatment process with him and the psychiatrist who saw her in college. In college, Susan continued the same up and down life: strong commitments, painful disappointment, and trouble with written assignments. Looking back, I think that whatever good both psychiatrists did came not from their insight-oriented therapy, but as a result of their relationship and support. Susan dropped out of college in her third year and not until her first psychotic breakdown 2 years later did we realize she was a schizophrenic. In looking back, I think I can see many earlier clues. Now, I question the ease with which pronouncements about "normal adolescence" are made. I also question the theory that early identification of schizophrenia is bad and necessarily leads to self-fulfilling prophecies. What about the wrong diagnosis of neuroses? In what other field is ignorance considered an advantage? A case could just as easily be made that early identification of children-at-risk could lead to early and more appropriate help. I particularly regret that I followed professional advice and distanced myself in her adolescence. Schizophrenics already feel too distant.

There are many other dynamics to explore. For example, was the bossy and argumentative nature of my brother and daughter a marker? When combined with shyness, could it be understood as first an attempt to get others to share their atypical perceptions, followed by hurt and angry withdrawal when they do not succeed?

Susan became catatonic at age 23, while living a nomadic life. She was hospitalized following the intervention of a policeman but released against the advice of the hospital and as a result of being informed of her rights by a patient advocate. We breathed a sigh of relief for the 3 days she was in the hospital, where we thought she would get help. We were informed of the diagnosis of schizophrenia at the same time she was released and were naturally shocked at the precipitate release. Susan wandered around various cities for another year and a half, living on the street and in shelters or crashing with friends. She was in touch every few weeks, sometimes phoning with fantastic accusations and threats. When she periodically came home, we tried to be caring and helpful and suggested she seek psychiatric help, but were afraid to push too hard. The one occasion on which we took a strong stand led to a violent outburst and flight. Her second catatonic episode mercifully came during a visit home, so we were prepared. This time we arranged for her commitment to a hospital nearby and had already chosen a psychiatrist. Medication got her moving again and helped her delusional thinking. She was more warm and outgoing than I had seen her in years. Susan went to a halfway house and enjoyed the group therapy and the contacts with the staff. We began to hope again. However, although she put up with the side effects of pacing for a month, she finally stopped her medication abruptly and lost all her gains. She took up a vagabond life again. Still we hoped we were ahead and that she would voluntarily seek help because of her positive experiences. We had not yet understood the awful grip of disorderd thinking. Instead, we were on the same merry-go-round with Susan: another involuntary commitment, another precipitous release with the "help" of a patient advocate, more self-destructive behavior, cries for help, and chaotic visits home.

We do not know if she will ever accept help or have a viable life. Yet, not knowing, we must somehow get on with our lives. We do this for ourselves, for each other, and for our son. As it is, he will carry an extra responsibility for the rest of his life, and we do not want to add to it. For my healthy son, who has had only the usual amount of human pain, and for my daughter, who lives in almost constant pain, I feel I must go on living as well as I can. One must search for choices and even pleasures if only on the margins and borders of the most pervasive tragedy. As I write, I realize I have been saying this to myself as well and have done so since as a child I witnessed my father's anguish about his schizophrenic brother and son.

7

Is Schizophrenia Contagious Psychologically?

Both critics and skeptics could complain that the masses of data presented and reviewed in the previous two chapters, which strongly implicate genetic factors of one kind or another in the etiology of schizophrenia, can be explained by environmental contagion. The vector, or agent, of that contagion could be some kind of within-family psychological phenomenon, such as the presence of a so-called schizophrenogenic mother or the affective style (expressed attitudes) or communication deviance between patently normal parents and between such parents and *some* of their children. The vector would have to work like a time bomb, not unlike the delayed action of slow viruses implicated in some physical diseases such as the neurological disease kuru, which even had a familial pattern of affection, since the age of onset of schizophrenia ranges from puberty to the middle 50s. As noted in Chapter 1, virus theories about the etiology of schizophrenia have a following among some serious scientists, including Edward Hare, Fuller Torrey, and Tim Crow. David King and Stephen Cooper (1989) note in their recent review that it is very important to separate viruses as causes of mental disorder from impaired immunological functioning arising from the stress of being mentally ill; neither psychiatric nurses nor the spouses of schizophrenics have elevated rates of schizophrenia.

Cogent arguments for the contributions of the family qua family to the causes of psychopathology can be traced back to the 1926 paper by the sociologist Ernest Burgess entitled "The family as a unit of interacting personalities." In 1928, no less a figure than Ludwig von Bertalanffy (1901–1972) — biochemist and philosopher of science extraordinaire — introduced an "organismic," or unified science, approach to the study of biological problems as a reaction to the overemphasis on a reductionistic approach in the life sciences at the time; his perspective was eventually called *general systems theory.* This shift in research paradigms was given impetus in the 1950s by Gregory Bateson, the Palo Alto cultural anthropologist, who had been very much influenced by Zen Buddhism and the thinking of Claude Shannon and Norbert Weiner (mathematicians/information specialists) on computer technology, cybernetics, and electronic communications. Bateson, with Don Jackson, Jay Haley, and John Weakland, drew on these ideas to formulate an influential "double-bind, communication deviance" theory about the causes of schizophrenia. In their view, schizophrenia was the result of pathogenic communication within families, the outcome of failure by children to integrate contradictory meanings in parental communications; by 1956, a new movement had been founded (see Theodore Jacob's *Family Interaction and Psychopathology: Theories, Methods, and Findings* [1986] for a comprehensive overview).

Designing research to test some of these ideas is very difficult, but the study of offspring born to schizophrenic mothers and fathers and then raised apart from them by adoptive parents selected for psychological health and not likely to overuse deviant styles of communication provides one very important avenue.

Adoption Strategies

When the parents who provide the genes to their children are not the same as the rearing parents, an opportunity exists to disentangle the genetic and the experiential factors that predispose an individual to the development of schizophrenia later in life. The traditional family studies of schizophrenia have already shown that the children of patients do indeed have a very high risk of developing the same disorder, at a rate some 13 times the general population rate. In adoption studies of schizophrenia, the adoptee of interest receives his or her genes (and prenatal environment) from birth parents, one of whom is schizophrenic, and his or her rearing environment and experiences from adoptive parents who have been carefully screened through interviews and hospital-record checks for psychological health and economic security. On the face of it, such an arrangement disentangles heredity from environment and permits a clearer examination of both classes of contributing forces. Ideally, well-designed adoption studies could either support the genetic interpretation of the observed familiality of schizophrenia or disprove such interpretations and point to psychological (or even infectious) contagion.

Four variations of the adoption research strategies exist. The first and most used looks at the grown-up, adopted-away offspring of birth parents (mothers and sometimes fathers) who are known to be or have been diagnosed as schizophrenic. The second design starts with grown-up adoptees who have been diagnosed as having schizophrenia and then proceeds to evaluate the current and retrospective psychiatric statuses of their biological *and* their adoptive families. The third variation, termed cross-fostering, which is very difficult to conduct, starts with the children of normal parents (or at least those without psychiatric records) who have been adopted into homes where an adoptive parent, apparently within normal limits psychologically at the time of adoption, later came to be diagnosed as schizophrenic. The final design, with no completed examples yet in hand, examines the adopted-away children of schizophrenics and their adoptive parents prospectively and longitudinally, as well as the physical and emotional environments to which they are exposed in their adoptive homes; once the yield of schizophrenic adoptees is sufficient, their families and experiences can be contrasted with those of the still-healthy adoptees so that the contributory and triggering events for schizophrenia, as well as the protective ones, can be specified.

Implicit in these strategies is the question of whether or not the presence of a schizophrenic parent is a relevant *environmental* factor in the development of his or her offspring's schizophrenia when it does occur. Does the simple presence of such a "pathogen" lead to the kinds of stresses, double binds, negatively expressed emotions, marital skews and schisms, communication deviances, and so on, that purportedly lead to the development of schizophrenia in those so exposed? If so, we would expect that removing children from such homes via adoption would result in a lower risk than the average of 13 percent usually observed among the offspring of schizophrenics reared at home (see Chapter 5). We know that about 87 percent of persons with a schizophrenic parent do not develop clinically overt cases of schizophrenia. From this observation alone it is clear that the presence of a sick parent is not sufficient as an environmental cause of schizophrenia; and, because only about 10 percent of schizophrenics have a psychotic mother or father, neither is such a parent necessary as a genetic cause of schizophrenia.

The only "experimentally" controlled environmental factor in adoption research strategies is the postnatal presence or absence of a schizophrenic parent who rears the child. Because we do not yet understand *which* aspects of such rearing experiences may be causal or contributory in the chain of events leading to schizophrenia, we cannot compare these aspects directly in the two kinds of household involved. The adoption results cannot be used to assess *general* environmental contributors to liability for developing schizophrenia and cannot be used to dismiss or belittle the importance of psychosocial or ecological factors, other than the one controlled factor, in a first episode of schizophrenia or in subsequent episodes.

One cautionary tale will illustrate the somewhat tortured explanation above. We now know that favism, a form of hemolytic (blood) anemia, is caused by a mutant gene on the X chromosome. It is quite common in the Mediterranean region, especially among Greeks and Italians. The disease does not show up until the carriers of the mutant gene begin to eat fava (broad) beans or are exposed to the pollen of the bean plants. If someone were to suspect that favism is caused by the exposure of sons (it is a sex-linked trait) to some unspecified psychological factor by their "favogenic" parents and then conducted an adoption study, the scientist could celebrate the confirmation of the favogenic hypothesis as long as the boys were adopted into a bean-free environment away from the Mediterranean. No anemias would appear among the adoptees, but anemias would continue to show up in a control group reared by their own bean-eating families. Both the gene and the bean are necessary in this instance, and any explanation focusing on only one causal factor would be wrong because it did not consider the interaction hypothesis.

Heston's Adoption Study in Oregon

The first published adoption study of schizophrenia was conducted by Leonard Heston (1966), then a young resident in psychiatry at the University of Oregon. He chased all over the United States and Canada in his Volkswagen, without a plush research grant, to accomplish a now classical result. Much of his success can be traced to the opening of doors for him by then-Senator Wayne Morse and to the selfless cooperation of many public servants. Heston interviewed the grown-up offspring of 47 women who had been diagnosed as having dementia praecox or schizo-phrenia in the Oregon state mental hospitals in the 1930s, such offspring having been placed in orphanages or with nonmaternal relatives during the first three days of life (as required by the law at that time). An appropriate control group of children was selected whose mothers had no record of mental illness, but who also were reared in foster or adoptive homes. The most common reasons for the separation in the control group were maternal death or desertion. The 50 control adoptees and the 47 "index" adoptees were all matched for sex, types of placement, and care, and were followed to an average age of 36. None of the fathers were found to be psychiatric patients in the Oregon state hospital system, but be-cause the babies were born to mothers while they were residents in the state hospital, we doubt that all the fathers were paragons of mental health and sound character. Heston told me that before beginning his adoption study, he expected his results to show that the risks to the children of schizophrenics reared by their own biological parents in the Kallmann studies had been exaggerated.

A summary of the findings and outcomes is provided in Table 11. Five of the 47 index adoptees grew up to be schizophrenic, as determined by the consensus of two blindfolded, independent clinicians plus Heston; none of the 50 controls was even considered to be psychotic. The 10.4 percent prevalence of schizophrenia among the adopted-away offspring of schizophrenics becomes 16.6 percent when the necessary age correc-tion is made. The rate is a bit higher in these offspring of severely ill mothers than the average of 13 percent reported in Figure 10 (page 96) for the age-corrected risk in children reared by their own schizophrenic parents. On the face of it, a transfer from exposure to a schizophrenic parent to some kind of "better" environment had no ameliorative effect on the rate at which the Heston sample became schizophrenic, but let us take a closer look at the circumstances.

The Oregon mothers all had severe cases of schizophrenia, whereas the schizophrenia of the parents of the offspring pooled for Figure 10 ranged from mild to severe, and we know from the family and twin studies already reviewed that greater severity predicts higher risks. What the fathers may have contributed genetically to the psychopathology ob-served in their offspring is uncertain, but the amount of "sociopathic personality" and other indications of personality disorder observed in the adoptees may well hint at a paternal contribution. What we need, and will provide shortly, is some kind of meta-analysis that examines the risk of schizophrenia in home-reared children of schizophrenics matched for severity of schizophrenia, coparent psychopathology, and postnatal expe-rience.

TABLE 11

Heston's Oregon adoption study: follow-up status of children separated
from schizophrenic birth mothers

	Born to Schizophrenic Mothers	Born to Normal Mothers
Number of adoptees	47	50
Male:female ratio	30:17	33:17
Average age	35.8	36.3
Schizoprenic offspring	5	0
Morbid risk (age corrected)	16.6%	0%
Sociopathic personality	9	2
Other psychiatric disorders	13	7
Mental retardation (only)	2	0
Average psychological adjustment rating (100 = best)	65.2	80.1
Average IQ	94.0	103.7
Average education (years)	11.6	13.4

Adapted from Heston (1966).

The Evidence from Denmark

Without replication of results, we do not have confidence-inspiring scientific data; we might be dealing only with chance or nongeneralizable
findings in the Heston adoption study. From Denmark, with its wonderful system of national registers (but under fire in 1989 and in danger of
extinction from citizens overly concerned about privacy), we have an
improvement in the design of the Heston adoption study by the addition
of schizophrenic fathers as parents of adoptees. The project was conducted by a team of NIMH (David Rosenthal, Seymour Kety, and Paul
Wender) and Danish (Joseph Welner, Fini Schulsinger, and Bjorn Jacobsen) scientists and has been available in part since the 1967 Dorado
Beach meeting. The researchers started with a national register of all

14,500 formal, nonfamilial adoptions in Denmark between 1924 and 1947, but used only the 5,500 children residing in the Greater Copenhagen area. The 10,000 known biological parents (25 percent of the fathers could not be specified by the mothers) were sought in the national psychiatric register mentioned in Chapter 5, which accurately identifies 95 percent of admissions since 1930 and, with less accuracy (Dupont, 1983), all admissions since 1910. At the time the NIMH-Danish team began, the registration system covering the current population of 5.5 million was not computerized; it has been since 1969 and now includes reports from 86 institutions with 12,000 beds. Scientists approved by governmental review boards can requisition and review all charts on every psychiatric patient, as long as they adhere to strict canons of privacy and confidentiality.

Initially, 69 biological mothers or fathers were judged to be psychotic from English summaries of their charts. Their children, ranging in age from 20 to 52, became the subjects of interest, along with their matched controls — adoptees who had unregistered ("normal") parents. At the time of the initial report, only 39 index and 47 control adoptees had been willing to undergo two days of intense evaluation with psychophysiological and psychological tests, including a three- to five-hour interview by a gifted, psychoanalytically trained Danish clinician, Joseph Welner. Some readers will want to examine the details of the study in the original reports and their extensions (Rosenthal et al., 1968, 1975; Lowing et al., 1983).

Only 27 of the parents of the index adoptees had "chronic schizophrenia," with the remainder carrying diagnoses of acute or borderline schizophrenia or manic-depressive psychosis. Of the 39 adopted children of the psychotics, 8 (21 percent) were blindly diagnosed as borderline schizophrenics or more disturbed ("hard spectrum"), and 13 (33 percent) altogether were considered to have either hard or soft spectrum disorders (i.e., single acute schizophrenic episodes or schizoid or paranoid personalities). The corresponding psychopathology figures for the control group were 2 percent hard and 15 percent soft, using the team's consensus clinical judgments. Note that these values are raw, non-age-corrected prevalences and therefore will be underestimates, because virtually all members of the sample (35/39) were still under age 45 (see Chapter 2) at the time they were evaluated.

When more objective DSM-III criteria, not in use at the time of the initial analyses, are imposed upon the Rosenthal adoption study, there should be an increment of reliability, certainly, but *no* guarantee of an increment in the validity of the results. Only one of the index adoptees had been hospitalized for schizophrenia, and only two others were given the diagnosis for unhospitalized episodes of what seemed to be schizophrenia; the rest of the observed psychopathology was less severe. The sample of parents of the adoptees was less chronically ill than was Heston's Oregon sample. Using DSM-III criteria, Patricia Lowing and colleagues (1983) at the NIMH reevaluated the original data collected on the Danish sample and reported that 39 parents (27 mothers and 12 fathers) met criteria for a diagnosis of schizophrenia, and only one of their offspring met the criteria, a prevalence of only 2.3 percent in the offspring; none of the controls met the criteria. Ten additional index and four control adoptees did meet criteria for schizotypal (a new category, close to earlier descriptions of borderline schizophrenia) or schizoid personalities. Thus, the new combined rate of hard and soft spectrum disorders is 28 percent in the index group and 10 percent in the control group, figures that are essentially in agreement with both Heston's study and the earlier reports by Rosenthal and colleagues, although the absolute levels of disorder differ.

The Danish research reduced some uncertainties left by the Oregon project. Fathers as well as mothers were the schizophrenic probands (baseline cases), so the likelihood was reduced that pregnancy-related prenatal and perinatal factors contributed significantly to the offspring risks (i.e., no schizophrenic fathers were pregnant!). Furthermore, the Danish schizophrenic parents included only a handful who had been obviously ill *before* their children were given up for adoption; thus, the adoption authorities and the adoptive parents could not engage in any self-fulfilling prophecies about the outcome of the children based upon knowledge about the biological parents' mental illnesses.

A Meta-Analysis of Offspring Risk

Let us now try to work out the simultaneous contributions of genetic and schizo-specific rearing factors to the risk of developing schizophrenia in the offspring of schizophrenics. The exercise has important implications for the theoretical battles surrounding schizophrenia research, as well as for the applied settings of courtrooms and custody hearings. As we noted earlier, the results of the adoption strategies reviewed so far make it appear that removal of a child from his or her schizophrenic parents has *no* effect on reducing the risk for schizophrenia. We did note that the prevalence of hospitalizable schizophrenia in the Danish study was only 2.3 percent, compared to the age-corrected average risk of 13 percent for own-home-reared children of schizophrenics (Figure 10); the adopted-away children from Oregon, however, still had an age-corrected risk of 16.6 percent. How can we reconcile these values with the common-sense notion that it must make *some* difference in risk if genetically vulnerable children are raised in physical and psychological chaos rather than in an average home? We can do it with a meta-analysis, using data that were not intended for such purposes originally, but that can be mustered into the service of a "nearly perfect" quasi-experimental design.

We observed in Chapter 5 that Kallmann's study, published in 1938, on the families of schizophrenics he had studied in the first third of this century, showed a very wide range of risk to offspring as a function of parental genetic and psychosocial characteristics, but especially to the severity of their illness as indicated by their Kraepelinian subtype. The overall risk of schizophrenia in the offspring of all his schizophrenics was 16 percent. That 16 percent, however, included some children with both parents schizophrenic (a 46 percent risk subgroup), some with a catatonic parent (a 22 percent risk subgroup), some with a paranoid schizophrenic parent (a 10 percent risk subgroup), and so forth. Kallmann also calculated risk broken down by whether or not the coparent was psychiatrically deviant (if not schizophrenic) and by whether the matings were illegitimate or not (thus identifying casual alliances between schizophrenic women and probably sociopathic males). We can therefore select subjects for our meta-analysis to match the characteristics of the parents in Heston's Oregon adoption study, where we deal *only* with schizophrenic mothers whose original diagnosis was Kraepelinian dementia praecox or schizophrenia, where the matings were largely illegitimate, and where the children were reared in much better circumstances than if they had stayed with their psychotic mothers.

Table 12 shows the risks that can now be calculated to answer the question about the effects of rearing by a schizophrenic parent on a child with the genetic predisposition to developing schizophrenia. Own-home rearing by a severely ill mother and illegitimacy resulted in an age-corrected risk for the offspring of 27 percent in the Berlin sample. For the adopted-away children with similar parental backgrounds in Oregon, the risk was only 17 percent; we say "only" because it could have been much higher. How much the risk was decreased by removing the child from a schizophrenogenic environment and rearing him or her in a foster or adoptive home is shown in the table as 37 percent; conversely, the risk would be increased by 59 percent if the child were to be left with his or her schizophrenic mother. Thus, although this meta-analysis requires some poetic license (we hope not so much as to make you reject it out of hand), it shows both the substantial genetic risk to the children of

TABLE 12

A meta-analysis of the strengths of familial environmental
factors in schizophrenia

Sample	Mating Type	Child Risk
HOME REARED		
(Kallmann)		
State hospital	Schizophrenic mother x "casual" male	27%
FOSTER/ADOPTION REARED		
(Heston)		
State hospital	Schizophrenic mother x "casual" male	17%

Therefore, % risk *increased* by exposure to schizophrenic mother	$=\dfrac{10}{17} = 59\%$	
Or, stated another way, % risk *decreased* by removal and fostering	$=\dfrac{10}{27} = 37\%$	

schizophrenics and the substantial incremental environmental risk if
they are left in their natal homes or, following parental divorce, are
exposed to contact with their schizophrenic parent. As we would expect
from a diathesis-stressor model, both the genotype and the postnatal
emotional climate are crucial to the observed outcomes.

Who Raised Schizophrenic Adoptees? The Kety Strategy

The various Danish registers already mentioned can be used in reverse
fashion from the Heston-Rosenthal strategy of starting with schizophrenic parents and following up their adopted-away offspring, as demonstrated in a now-classic study by Seymour Kety and his team of
American and Danish colleagues. Adoptees who grew up to be schizophrenic and control adoptees who had healthy outcomes became the
starting point of the investigation, and their biological and adoptive
families became the subjects of further scrutiny. Danish psychiatrists

conducted field interviews and searched the national register for both biological and adoptive relatives of the schizophrenics who had significant psychopathology, especially schizophrenia and putative schizophrenic spectrum disorders. Results from both the total nationwide sample of 14,500 adoptees and a Greater Copenhagen sample of 5,500 of these same adoptees have been reported in the scientific literature, and the conclusions are essentially in agreement (Kety, 1988). Thirty-three adoptees who, when they grew up, had become schizophrenic were identified from the Copenhagen sample, and 41 more were added when the search was extended nationwide — a tribute to the perseverance of the investigators and to the usefulness of thorough national statistics for mental health.

Only the biological relatives of the schizophrenic adoptees showed a significant concentration of schizophrenia and schizophrenia spectrum disorders; the adoptive relatives, those who had reared and lived with the schizophrenics, did not differ from the rates in the control populations for such conditions. For the Greater Copenhagen sample, 21.4 percent (37/173) of the biological relatives had spectrum disorders (6.4 percent had definite schizophrenia), and for the adoptive relatives, 5.4 percent (4/74) had spectrum disorders (1.4 percent had definite schizophrenia). Clearly, it was something the schizophrenic adoptees had inherited from their biological schizophrenic parents that caused their mental illness, because the adoptive families that had actually raised the schizophrenic adoptees had much lower rates of schizophrenia.

Another major piece of evidence about the contributors to schizophrenia comes from the analyses conducted by Kety and his colleagues on the 104 biological half siblings of the adoptees who grew up to be schizophrenic. Among the 63 *paternal* half siblings of schizophrenic adoptees, 12.7 percent had definite schizophrenia; while among 41 *maternal* half siblings, the rate of schizophrenia was 4.9 percent. Because the former did not share the same uterus or early mothering experiences with a schizophrenic and the latter did, the results go in the opposite direction from that predicted by the hypothesis that obstetrical/neonatal events make important contributions to the eventual development of schizophrenia.

In 1984, two young American psychiatrists, Kenneth Kendler and Allen Gruenberg, took on the formidable task of reviewing the translated Danish material for all the adoptees and those of their relatives who had been interviewed in order to derive new, "more objective" clinical diagnoses based upon the criteria enumerated in DSM-III. The project was necessitated by the growing disquiet with the past use of clinical criteria to support the canon about the role of genetic factors in schizophrenia. A summary of their findings for the Greater Copenhagen sample is shown in Table 13. Most important, the overall occurrence of schizophrenia or schizophrenia spectrum disorders is still significantly higher in the biological relatives of schizophrenic adoptees, relatives who *did not rear them*, than in the adoptive relatives with whom and by whom they *were* reared. Thus, the original interpretation was sustained and strengthened: Genes from schizophrenic parents are more important in producing schizophrenia than is rearing by those biological parents.

Between the original clinical investigation in the 1960s and the application of objective criteria in the 1980s, the schizophrenia spectrum was redefined and refined to include not only DSM-III schizophrenia, but also schizoaffective psychosis — mainly schizophrenia — schizotypal personality, and paranoid personality. Schizophreniform disorders (as defined in DSM-III) were excluded from the spectrum in this reanalysis. Casual readers need not be overly concerned with such details, but novice researchers must track them obsessively lest they make false claims of confirmation or refutation in future studies that build on these foundations. The objective criteria of DSM-III shrank the original 34 index schizophrenic adoptees to 19 schizophrenic spectrum cases, but a further 5 who had been called schizophreniform had been excluded under the rules. Among the 69 first- and second-degree (parents and half siblings) biological relatives of the spectrum adoptees (data line 2 of Table 13), 22 percent were blindly diagnosed as having spectrum disorder when defined in this fashion, versus only 2 percent of the control's relatives, a difference so reliable statistically that you could bet the farm on it.

For the first time, the age-corrected lifetime morbid risks for schizophrenia per se can be calculated in an adoption study for comparison with the data in Figure 10. For the first-degree relatives, the risk is 10.5 ± 9.9 percent (the standard margin of error statistic) and for second-degree relatives, it is 6.7 ± 6.5 percent, close enough to both estimates from ordinary family studies to say that they are all in agreement with one another.

Once again, using this unique variation of the adoption strategy, we can reasonably support the hypothesis that schizophrenia appears to occur through an interaction or a coaction of some kind of genetic susceptibility with some kinds of environmental stressors, prenatal and/or postnatal. The stressor need not be an environment including a "spectrum person," as none of the adoptive parents had such diagnoses. The genetic factors appear to be reasonably specific — the schizophrenic biological parents of the adoptees did not transmit any increased risks for major affective disorders or for delusional disorder to them. If anything, the adoption study results give too much weight to genetic factors in

TABLE 13

Kety's schizophrenic and control adoptees: frequency of spectrum disorders in their biological relatives using DSM-III diagnoses

| Adoptees | | Biological Relatives | | | | | | | |
| | | 1st degree | | | | 2nd degree | | | |
Diagnosis	N	N	Schizophrenia	Schizotypal personality disorder	Paranoid personality disorder	N	Schizophrenia	Schizotypal personality disorder	Paranoid personality disorder
Schizophrenia	13	10	1	2	1	25	1	3	1
Schizophrenia spectrum*	19	17	1	3	2	52	2	6	1
Screened controls	24	31	0	0	0	60	0	0	0
All controls	34	47	0	1	0	90	0	1	1

After Kendler and Gruenberg (1984).
*Schizophrenic spectrum (13 + 6) includes schizophrenia, schizotypal personality disorders, schizoaffective-schizophrenic type, and paranoid personality disorder.

causing schizophrenia; the results must be interpreted in light of the meta-analysis above, which tried to control adequately for the role of severity of illness in the biological parents to show that good adoptive-home rearing could indeed make a significant difference in protecting vulnerable offspring from developing schizophrenia and in reducing their risk considerably — but not eliminating it.

Wender's Cross-Fostering Research

So far in the history of psychiatric genetics, only the Danish record system permits the carrying out of the unusual design followed by Paul Wender (1974) and the Danish-American team described above. Through no fault of their own or of the conscientious adoption authorities, some children of apparently normal parents will be placed for adoption with normal-at-the-time families where the adoptive mother or father later develops a schizophrenic disorder. Such cross-fostering occurred 28 times among the 5,500 adoptees in the Greater Copenhagen sample. The risk of schizophrenia spectrum disorders in such a unique sample provides a very direct test of the influence of so-called schizophrenogenic mothers and related hypotheses. If any of these hypotheses are valid, a high proportion of the children of normals reared by schizophrenics should develop conspicuous psychopathology. Of the 28 cross-fostered subjects, 3 (10.7 percent) were considered to have probable borderline schizophrenia or more severe schizophrenia-like pathology. After purifying the sample of 28 cross-fostered children by removing those who were deviant before adoption (e.g., those who had contracted congenital syphillis) or who had a birth parent with a psychiatric diagnosis other than schizophrenia, the sample was reduced to 21 and the rate of spectrum disorders among them was reduced to 4.8 percent. How does this rate compare with that of the children of schizophrenics adopted into the homes of normals (the Rosenthal design above)? Wender reports a rate of 18.8 percent of hard spectrum diagnoses using the same criteria that yielded the much lower rate of 4.8 percent. However, only 9 of the 28 original adoptive parents who became schizophrenic were diagnosed as having chronic schizophrenia; the remainder suffered milder illnesses. These facts weaken the test somewhat but, on balance, it would seem that in the absence of genetic predisposition (as inferred from the fact that the children were born to normal parents), the exposure to "schizophrenic parenting" is not a powerful factor. On the other hand, given the genetic predisposition (as inferred from the fact that at least one of the birth parents was schizophrenic), rearing by normal, adoptive parents provides limited protection from developing schizophrenia. Refer again to the meta-analysis above for complications to these simplified and tentative summary statements. No doubt, imposing DSM-III criteria on the cross-fostering data would weaken but not eliminate the demonstration of results that are complementary to the other two Danish adoption studies and the Oregon study.

Adoption Strategies in Finland

The adoption studies described above could only be retrospective and they concentrated on the final outcome statuses of the adoptees. Furthermore, the actual yields of definite schizophrenics among the various critical groups in the adoption studies in Oregon and Denmark were too small to be definitive and confidence-inspiring. An analytical, or fine-grained, approach to the actual child-rearing practices and familial interpersonal styles collected prospectively with the adoption strategy would remove much of the remaining ambiguity about how the adoptees actually got to be the way they turned out—sick or well. Such a difficult undertaking is now under way in Finland, led by the well-known Finnish psychiatrist Pekka Tienari (see the discussion of his study of twins in Chapter 6) and the eminent American psychiatrist and psychologist Lyman Wynne, together with a team of Finnish colleagues. These leading scientists are predisposed to psychodynamic and family-systems theories about mental illness, although they cautiously accept the necessary roles played by genetic factors. Wynne and the clinical psychologist Margaret Singer were early proponents (1965) of an epigenetic view of the origins of schizophrenia, proposing important roles for both predisposing genetic factors and parental communication deviance.

From a starting pool of 20,000 schizophrenic and paranoid psychotic women in the nationwide Finnish sample of women hospitalized from 1960 to 1980, 171 have been confirmed as having adopted away their 184 offspring to nonrelatives before the children were 4 years old. Appropriate controls with normal mothers were also selected and matched with the index offspring on a case-by-case basis for important variables. Very preliminary results for only the retrospective part of the study, in which the offspring who were age 16 or older were looked at, confirm the American and Danish outcome results. For the first 128 index and control offspring studied so far, 10 psychotics were found among the adopted-away offspring of schizophrenics (broadly defined; 8/93 when the strict RDC criteria were used), and only one psychotic was found among the adopted-away offspring of normal controls. Age correction would be premature so early in the project.

What is new and exciting, in addition to confirmation of the earlier results even by investigators not genetically inclined, is the preliminary observation that the mental health statuses of the *adoptive* families could be differentiated. The data came from clinical ratings of communication deviance and ratings of the interpersonal dynamics within couples, including the use of Rorschach Ink Blots. Five of the ten psychotic index adoptees and the one control case were reared in families that received the worst clinical rating, "severely disturbed"; a further three psychotic adoptees came from adoptive families rated "moderately disturbed." Of course, a number of other adoptees reared in similarly rated adoptive families did not become psychotic.

The diagnoses of the 10 psychotics include 6 schizophrenics, 3 paranoid psychotics, and 1 manic-depressive. If paranoid psychosis, a.k.a. delusional disorder, is not genetically related to schizophrenia (Schanda et al., 1983; Kendler et al., 1987), the initial findings would still confirm the earlier studies, but interpretations of the data would become more complex. When soft data from this project and its prospective component become "hardened" by blind ratings of the audiotapes (initially, only one interviewer was permitted into the homes of the adoptive families) and are available for peer criticism, they could support the diathesis-stressor model for the origins of schizophrenia that we have been emphasizing throughout this book.

Continuing uncertainty about whether schizophrenics-to-be cause the observed disturbance of family communications or disturbed family communications lead to schizophrenia will need resolution from the prospective part of the Finnish adoption studies. Such a resolution could also provide the rationale for specifying the relevant psychosocial and ecological stressors in the diathesis-stressor model for schizophrenia and thus provide the long-sought information needed for rational programs of intervention and prevention.

Overview of Adoption Results

The broad genes-plus-environment hypothesis for explaining the cause of schizophrenia is strengthened by the adoption studies. They show that sharing an environment with a schizophrenic parent or another schizophrenic generally does not account for the familiality of cases. One of the favorite notions before 1966 (when Heston's adoption study and Gottesman's and Shields's twin study based on objective diagnoses were published) was that genes inherited from a schizophrenic did predispose a person to schizophrenia, but that the transmitter of the genes who also raised the person transmitted a schizophrenogenic environment as well. Both the necessity and the sufficiency of the specific kinds of schizophrenogenic environments provided by schizophrenic parents have been weakened by the adoption results. Recall that almost 90 percent of schizophrenics do not have schizophrenic parents.

It is clear, though, from the Finnish work in progress and work to be reviewed in the next chapter on other intrafamilial influences (expressed emotion, affective style, etc.), that such influences have not been ruled out as part of the complex causal chain leading either to the development of schizophrenia or to changes in its course. But such factors are not specific to the parents of schizophrenics, and those parents must not bear that cross any longer. The causal factors may well be like those in our story about the fava bean and hemolytic anemia; the environmental factors are quite common in some environments, but result in a disease or disorder *only* when they interact with specially predisposed genotypes.

8

The Role of Psychosocial and Environmental Stressors

Genetic strategies and inferences about the etiology, course, and outcome of schizophrenia are much easier to devise than are those required to detect psychosocial and physical-environmental factors. Grist for the environmental research mill is elusive even when invoking the important, if exotic, variations on the adoption strategy discussed in the previous chapter. Reliably tracing events that occurred during the two or three decades of living before the onset of schizophrenia, let alone during prenatal development, is next to impossible in the absence of such aids as national case registers for mental illnesses, compulsively kept "baby books," and libraries of home videotapes.

Even more troublesome is the probability that environmental effects do not impact a preschizophrenic's life as single events, but instead act cumulatively; thus, the casual observer sees and records only *the* straw that allegedly broke the camel's back. How one event affects an individual depends not only on how it is perceived by that individual and on what events have preceded it, but also on when the events happened and on individual differences in the person's vulnerability at each of those times. Such a series of evolving transactions has been termed *epigenetic* by Margaret Singer and Lyman Wynne, as noted in Chapter 7, and has been expanded by Gottesman and Shields (1982) to include changes over time in a person's constitution (the sum total of his or her physical self) associated with the turning on and off of genes in response to changes in external and/or internal environments.

It would be a boon to therapeutic intervention, to rehabilitation, and to prevention if systematic clinical research could eventually uncover a few major factors within the context of the psychosocial or ecological environment that cause or contribute to the onset of schizophrenia, maintain an episode, and influence relapses and recoveries. More likely, we may eventually begin to see that numerous minor factors come together (concatenate) in ways that trigger or worsen a schizophrenic episode. We may also learn that certain other factors experienced alone or in combination either prevent or reduce the duration of severity of a psychotic episode. From such empirical observations on validly diagnosed schizophrenics, we can build our multifactorial genetic-plus-environmental models using the data reviewed in the previous chapters to guide us.

Even under the best circumstances, the identification of universal, important environmental factors in the epigenesis of a disease or disorder is difficult. In earlier chapters, it was shown how such macrolevel factors such as social class, nationality, degree of industrialization, and the presence of a schizophrenic parent could be discounted as universal factors. Death from lung cancer provides a case study in the identification of a risk factor that contributes to the development of a disease. The death rate from lung cancer among *nonsmoking* British male physicians is 7 per 100,000 men. Among physicians who smoke 20 or more cigarettes a day, the death rate is 139 per 100,000—a 20-fold increase in relative risk. Since we have controlled for occupation and social class, can we safely conclude that cigarette smoking invariably *causes* lung cancer? The vast majority of smokers do not die of lung cancer, but all unbiased observers, especially the Surgeon General and the tax and health insurance premium payers, would conclude that a major risk factor—tobacco consumption—had been identified, one with immediate implications for prevention. By inference, individual differences in vulnerability to lung cancer must also be involved. One could consider the relative risks for schizophrenia in the offspring and identical cotwins of schizophrenics (see Chapter 6) of 13-fold and 48-fold, respectively, as implicating important risk factors, but this time these factors would be genetic. Again, since the majority of such offspring and cotwins are *not* schizophrenic, nongenetic contributors to the development of schizophrenia are implicated. It is obvious that we have just restated the diathesis-stressor formulation, which calls attention to genetic assets and liabilities as well as to environmental assets and liabilities (see Figure 8, page 90). With respect to the relative importance or power of such contributors, it would appear that sharing 100 percent of the genes of a schizophrenic is more than twice as important in causing schizophrenia as smoking a pack a day is in causing death from lung cancer.

Acute and Persistent Life Events

Progress has been slow in amassing a list of stressors in the diathesis-stressor context that will be found in association with the onset or the relapse of an episode of schizophrenia. It is easy to conclude that the putative life events are idiosyncratic; that is, what may be a stressor for one preschizophrenic rolls off the back of the next like water on the proverbial duck, while for someone without a schizophrenic diathesis, that very same stressor may precipitate anxiety or depression or go unnoticed.

Common sense would lead one to believe that a life-threatening event such as fierce wartime combat would surely be among the stressors associated with the precipitation of a predisposition to schizophrenia. Common sense, in this case, would be incorrect. During the first two months of the Normandy offensive, for example, the Army set up "exhaustion centers" near the beachheads to handle the expected influx of neuropsychiatric casualties. (Combat exhaustion had been called "shell shock" in Word War I.) Philip S. Wagner was the chief psychiatrist for the unit that was set up 12 miles inland from Omaha Beach on D-Day plus 14. After the war (1946), he reported on his experiences. Within the first 48 hours of center activity, 275 men were admitted, many just waiting at clearing stations for the doors to open; within 8 weeks, 5,203 neuropsychiatric casualties were treated at just this one center, which served four to eight divisions over that interval. The remarkable fact is that only 66 of these casualties received a diagnosis of schizophrenia — a mere 1.3 percent — and only 154 altogether were considered to have any kind of psychosis. The vast majority of the casualties were diagnosed as either anxiety neuroses or "inadequate personalities." Similar results were reported for the British Army at the invasion of Dunkirk (Slater, 1943). It is clear that catastrophic life-threatening stress, when in military men enduring combat or in civilians enduring the blitz or internment in concentration camps, can lead to serious psychiatric problems that today would be called posttraumatic stress disorders, but such casualties apparently do not become schizophrenics very often. Perhaps the operative stressors for schizophrenia are more subtle.

H. Steinberg and J. Durell (1968) examined volunteers and draftees under peacetime conditions to test the idea that the necessity of adapting to basic military training by young men separated from their families for the first time and treated with the impersonal harshness that is routine could precipitate schizophrenia. These men had all been screened to ensure mental health. Hospitalization rates for schizophrenia were markedly elevated during the first month of service, six times higher than the rate during the second year of service. The effects during the first month were more marked on draftees than on volunteers. A careful review of records showed that these cases of schizophrenia were not simply the result of psychopathology being overlooked at induction screening. The researchers concluded that "The hypothesis with which the data are consistent is that the emotional stress associated with military social adaptation was effective in inducing schizophrenic symptoms" (p. 1103).

Clinical experiences with a few patients often may lead to the strong impression that certain life experiences or events had a causal role in a particular patient's development of schizophrenia. William Schofield, a clinical psychologist at the University of Minnesota School of Medicine, and Lucy Balian, then a senior medical student, put such impressions to a direct test by conducting the same kind of penetrating interviews with 150 psychiatrically normal persons as is routinely conducted with psychiatric patients. The subjects were patients with serious physical illnesses who had not been referred to the hospital for psychological reasons. They were matched with 178 schizophrenics in the same university teaching hospital on a host of demographic variables (e.g., social class, education, marital status). It is quite an eye-opener to see how similar the histories of such divergent groups were. Table 14 highlights some of the findings with respect to interparental relationships in the rearing home, maternal emotional attitudes, and such home conditions as poverty, divorce, and early parental death.

Of the 35 major aspects of early history and adjustment covered by the interviews, one-third did not show a significant difference between the families of schizophrenics and those of normals. For the 22 variables on which a reliable difference did emerge, 5 (23 percent) showed the normals to have experienced *more* psychosocial disadvantages than the psychotics. Although maternal affection was less characteristic of the rearing of schizophrenics, two-thirds of them still reported a positive experience. Furthermore, given the retrospective nature of the research, some of the observed pathological findings, such as maternal overprotection, may have been the result rather than the cause of early symptoms of mental disorder. The Schofield and Balian study calls into serious question much of the received wisdom about the schizophrenicity of many antecedent events, but enough positive findings remain to suggest that

TABLE 14

Interparental and maternal relationships and home conditions in the
early histories of normals and schizophrenics

Relationship	Normals (N ≥ 144)	Schizophrenics (N ≥ 101)	Is Difference Statistically Significant?
INTERPARENTAL			
Affection	75.7%	76.2%	No
Ambivalence	6.3	5.9	No
Indifference	0.7	1.9	No
Hostility	17.4	15.8	No
MATERNAL			
Affection	81.3%	64.8%	Yes
Ambivalence	8.0	2.3	
Indifference	2.7	1.5	
Rejection	6.0	6.2	Yes
Overprotection	0.7	13.2	
Domination	0.7	10.9	
Neglect	0.7	0.7	
HOME CONDITIONS			
Poverty	20.7%	9.0%	Yes
Alcoholism	6.0	6.8	No
Invalid	12.7	0.6	Yes
Divorce	6.0	1.6	No
Separation	2.7	1.1	No
Death of parent	14.7	10.7	No

Adapted from Schofield and Balian (1959).

further research is in order. The fact that almost one-fourth of normals
reported "traumatic experiences" without disastrous outcomes draws
attention to the need to study a person's assets in addition to the usual
focus on liabilities.

After a lifetime of research that evaluated what happens to a schizophrenic in the weeks and months before first onset or relapse, Bruce
Dohrenwend, an eminent social psychologist at Columbia University
specializing in epidemiology and psychopathology, reached the following

conclusion in 1987: "Recent environmentally induced stress from life events does not appear to be of primary importance as a risk factor for episodes of schizophrenia. Accepted at face value, the results suggest that in the search for environmental causes we give greater emphasis to class-related socialization experiences in adverse family and other early social environments. Relative to such factors, recent stressful life events and networks appear to be of secondary and derivative, though not necessarily neglible, importance" (p. 292).

The "New Look" in Family Influences

In the quotation above, Dohrenwend is asking for an emphasis on family variables other than the ones emphasized for so many years by the first generation of psychoanalytically oriented family-systems researchers mentioned in the previous chapter, such as Bateson, Jackson, or Lidz. Starting in the mid-1960s, a conceptual revolution took place among many family researchers interested in nongenetic and nonbiological aspects of schizophrenia as a consequence of the follow-up studies of discharged schizophrenic patients conducted by the Medical Research Council (MRC) Social Psychiatry Unit at the Institute of Psychiatry in London under George Brown, a sociologist with training in anthropology, and his psychiatric colleagues, John Wing, James Birley, and Michael Rutter. They had observed that discharged schizophrenic patients were more likely to have early relapses if they returned to close ties with their parents or a spouse. The concept of disruptive *emotional expression* (EE) evolved from a composite of the putative influence on relapse of high emotional overinvolvement with the patient, critical comments directed at them, and overall hostility directed at them. Because of the follow-up studies, the EE school of thought and its heirs focused not on the factors responsible for the *first* appearance of schizophrenia, but rather, given that someone has already developed schizophrenia, on what factors appear to make it worse (cause relapse) or better (prevent relapse). By separating questions about etiology from those about influences on the course of the illness, the "cold war" that had developed between biogenetic and psychosocial-family systems researchers and clinicians (see Rosenthal and Kety, 1968, and Wynne, Cromwell, and Matthysse, 1978) can be declared over, with both sides victorious and the field in need of its own "Marshall Plan." As Julian Leff, currently assistant director of the MRC Social Psychiatry Unit, and Christine Vaughn, a psychologist with expertise in the evaluation of schizophrenics' families, note in their definitive book about EE, "The contrast between these two questions and their implications explains why the line of reasoning pursued . . . has been so much more productive to date than the work on the etiology of schizophrenia [by the earlier generation of family theorists]" (1985, p. 3). Some family systems researchers prefer to term the putative operating variables *high communication deviance* or *negative affective style*. More studies are needed before the degree of independence and overlap among the elements of expressed emotion, communication deviance, and affective style becomes clear. Because most work has been done on EE, it is emphasized as a prototype for psychosocial stressors in this chapter. A concise and informed review of these and other social influences on schizophrenia may be found in the paper by Paul Bebbington and Liz Kuiper (1988) at the Maudsley Hospital.

Brown, recounting the history of the discovery of EE, confesses that he was not aware of the work of Faris and Dunham or Hollingshead and Redlich (see Chapter 4) and had avoided reading psychiatric accounts of the etiology of schizophrenia by his colleagues at the Maudsley Hospital with well-developed diathesis-stressor views; that is, genetic predispositions augmented and/or released by environmental stressors (such as those of Eliot Slater and Martin Roth). Nonetheless, he adds, "I never questioned the broad relevance of the diagnostic label of schizophrenia, nor even the presence of an important genetic component, but [rather] the interpretations that had been placed upon the diagnostic label" (1985, p. 11).

A very impressive set of results was published by Vaughn and Leff in 1976, combining their data with those of Brown, Birley, and Wing (1972), by simultaneously exploring the effects on the relapse rates of discharged schizophrenics of three variables: continuing versus stopping medication (antipsychotic drugs such as phenothiazines); low versus high levels of EE in the home (assessed from a structured interview with the relatives *about* the patient); and, for the high EE patients, the number of hours of face-to-face contact between the patient and relatives. Earlier results had shown that in the nine months following discharge from hospital, schizophrenics going to high EE households had a 58 percent relapse rate, while those lucky enough to be discharged to a low EE family had a rate of only 16 percent.

The global variable of EE was derived from interview items with patients' relatives that sample five clusters: number of critical comments (six or more is called bad), emotional overinvolvement, hostility, warmth, and positive remarks. Some of the poor reception by the families of the mentally ill to the EE program of research both in the United Kingdom and the United States stems from the emphasis on the first three elements, with their reminders about "blaming the parents" for the schizophrenias of their offspring; a public relations effort should restore good-will with the families once they are made aware of the positive elements. The negative elements are very often elicited by the infuriating demands and outrageous behavior of many schizophrenics living with, or even in telephone contact with, their relatives; only the rare "Mother Theresas" can maintain their composure and equanimity under such chronically exhausting conditions.

Figure 11 diagrams in flowchart fashion the results of the more comprehensive design controlling for medication, EE, and exposure duration. By far the worst relapse rate was experienced by the patients from the worst of all worlds: 92 percent relapsed when they stopped their medication, lived in a high EE environment, and did so for more than 35 hours per week of face-to-face contact. Patients discharged to low EE homes had only a 13 percent relapse rate, and the taking of medication (for at least eight of the nine months) did not appear to be relevant as a buffer in this subsample. Medication seems to have been especially effective in preventing relapse for those schizophrenics going into high EE homes.

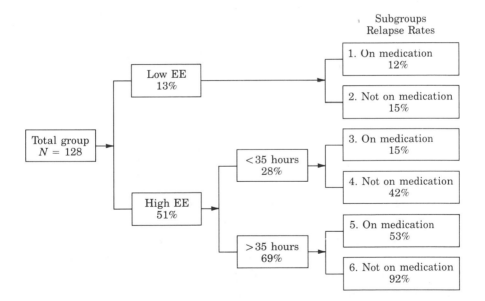

FIGURE 11.   Rates of relapse from recovery back to schizophrenic episodes in a sample of 128 British patients followed for nine months after hospital discharge, as a function of maintaining antipsychotic medication or not, exposure to EE (negative expressed emotion), and weekly duration of exposure to high levels of EE. Adapted from Vaughn and Leff (1976).

In a continuation of these efforts and an exciting expansion into interventions based on these data, the MRC group in London studied the outcome of behavioral intervention (a common sense approach to impart knowledge about the facts of schizophrenia and the everyday management of stress) with families of schizophrenics at high risk of relapse. They selected patients returning to emotionally charged homes. All patients took neuroleptic drugs, but 12 families received routine outpatient care and 12 received careful attention in the form of an educational program — a family discussion group and family sessions for relatives and patients. At follow-up after nine months, 50 percent of the control group, but only 9 percent of the experimental group, suffered relapse. The researchers found that in families in which social intervention was effective (73 percent of the 12), there had been no relapse. Self-selection for cooperative (perhaps the healthier) families is obviously an important factor to consider in evaluating the reported successes of family intervention and rehabilitation; many schizophrenics are not located within a family or other support network.

Research into cost-effective ways to rehabilitate schizophrenic patients and to reduce relapse rates has mushroomed in the wake of the positive findings reported above. It should be noted that preventing relapse is not the same as restoring the patient to full psychosocial and occupational functioning, an extremely difficult demand given the pervasiveness of "negative symptoms" in nonrelapsed patients maintained on medication and clearly helped by a reduction from high to low EE. The reduction has been mediated by various forms of family therapy focused on reducing the number of critical comments and the amount of face-to-face contact discharged patients have with overinvolved relatives.

In a masterful overview of these rehabilitative efforts, Angus Strachan has chronicled the successes of the MRC offshoot groups led by Michael Goldstein, Ian Falloon, and Robert Liberman at the University of California at Los Angeles; Gerard Hogarty at the University of Pittsburgh; and the continuing work of Leff and his colleagues. Family education, family behavioral therapy, and family groups (less demanding of professional time than family therapy — like a cooking lesson from Julia Childs) have all been used in various ways for patients who have been maintained on standard or reduced (Hogarty et al., 1988) doses of antipsychotic medication. Figure 12 shows the relapse rates for drug-maintained schizophrenics with and without some form of family intervention. As a frame of reference for evaluating such data, it must be kept in mind that even without any medication, about 20 percent of schizophrenics will have a "spontaneous remission" of an episode of illness, but there is no known way to identify those individuals specifically. A failure

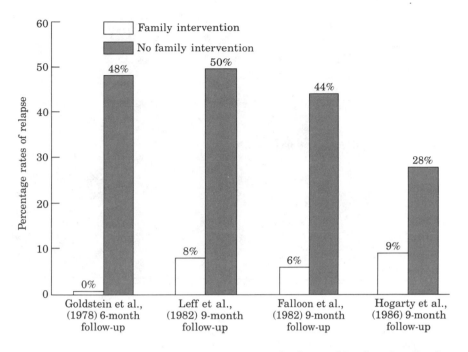

FIGURE 12. Rates of relapse from recovery back to schizophrenic episodes from four studies in the United States and England examining the protective effects of family therapy interventions versus no such interventions in schizophrenics who were continuously maintained on antipsychotic medications. Adapted from Strachan (1986).

to reduce either EE or the relapse rate among schizophrenics has been reported from West Germany; however, the kind of group therapy implemented was quite psychodynamic, which may have led to untherapeutic overstimulation, unlike the interventions represented in Figure 12.

One important finding that has emerged from efforts to identify the "active ingredient" in such interventions is that a patient's psychophysiological activity, an indication of emotional excitement or arousal, decreases when a low EE relative is present, but stays elevated when a high EE relative is present (Tarrier et al., 1988; Leff et al., 1989).

One scenario of how drug therapy, high EE, and life events may concatenate is illustrated in the life of one schizophrenic by the graph in Figure 13. Two thresholds are shown, beyond which symptoms appear: a higher one when the patient is taking medication and a lower one (i.e., more susceptible to relapse) when the patient is not taking the prescription. The chronic stress of living with a high EE relative is assumed to be additive with any acute stresses from intermittent stressful life events.

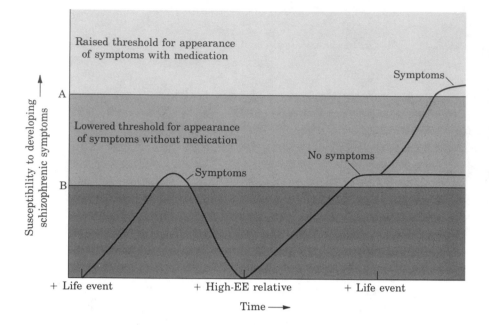

FIGURE 13.   A graphic representation of a schizophrenic patient's cumulative susceptibility to environmental stressors with medication (section A) and without medication (section B). Adapted from Leff and Vaughn (1985).

Even though the threshold has been raised by taking medication, relapse occurs when the patient must cope with both intensive face-to-face contact with a high EE relative *and* a concurrent stressful life event.

In their summary of the literature on EE, Leff and Vaugh conclude, "These findings indicate that the emotional attitudes measured represent nonspecific stress factors in the pathogenesis of schizophrenia. The specificity must lie in the schizophrenic patient's reaction to stress; we believe in the nature of the biologically determined vulnerability" (1985, p. 184). Findings from studying EE in the relatives of neurotic depressives, manic-depressives, anorexics, and obese women have all shown high levels of EE within the families of individuals suffering from those conditions as well.

It is time to smoke the peace pipe.

Obstetrical Complications and Other Somatic Risks

Many other environmental risk factors have been examined in the literature on schizophrenia besides the ones discussed so far in this book. Research focused on obstetrical complications as predisposing factors seems especially promising, since birth injuries have been repeatedly implicated in retrospective studies of schizophrenic adults, as well as in the children born to schizophrenic women. Recent applications of the newer brain-imaging techniques such as CT scanning have revealed that birth complications in schizophrenic twins (Adrianne Reveley and colleagues in London) and in the children of schizophrenics who grow up to be psychiatrically deviant (Fini Schulsinger and colleagues in Copenhagen) are associated with abnormal enlargements of the lateral cerebral ventricles. Inferences about insults to the brain are on much firmer ground nowadays, and such insults provide evidence for the multifactorial approach to the causes of schizophrenia. We spoke earlier, in Chapter 6, of an inconclusive correlation between lowered birth-weight and later schizophrenia in discordant twin pairs, but that is only one obstetrical factor among some 60 that have been identified. *Obstetrical complications* (called OCs in the trade) is a term used to cover pregnancy complications, birth complications, and complications occurring within four weeks of birth. Studies have looked at births to current schizophrenic mothers or wives of schizophrenics, retrospectively at the births of current schizophrenics, and even prospectively to see if anything can be noted at birth that later becomes relevant to schizophrenia. Despite all this effort, the available evidence is inconclusive and often contradictory.

A thorough and provocative review of this whole area by the American-Swedish psychologist Thomas McNeil and the late Swedish psychiatrist Lennart Kaij shows that an earlier focus on birth weight was too specific. A broader view of obstetrical complications supports the conclusion that in studies of schizophrenic identical twins, the one with the most OCs is more often the one affected or, if both are schizophrenic, the one more severely affected. Twins are more liable to obstetrical complications than are singletons and identical twins are more at risk than are fraternal twins. Moreover, males are at greater risk than are females. Thus, if obstetrical complications had a *specific* causal role in schizophrenia and were not just risk-increasing contributors in a network of causes added to a genetic predisposition, we would have certain expectations: Schizophrenia should be more prevalent in identical twin males than in females, in identical twins than in fraternal twins, in males than in females, and so on. No such predictions have been confirmed for schizophrenia, although they have been for mental retardation.

Even with epilepsy and mental retardation—conditions known to depend on the physical integrity of the brain—it has not been easy to implicate OCs and other biological factors such as viral infections and anoxia. When, as with schizophrenia, the condition isn't detectable for 15 to 50 years after the event, the problem seemed insurmountable until the arrival of the newer brain-imaging (CT, MRI) and brain-functioning

(PET, regional blood flow, improved EEG) techniques. Although McNeil and Kaij have reservations about obstetrical-complication research, they believe these complications to be independent stress factors that may interact with genetic factors. The task of further research is to specify which complications, how important they may be, and in what proportion of cases they are operative.

Seek Simplicity and Distrust It

The message in this aphorism from the writings of Alfred North White-head, mathematician and philosopher, while describing the concept of nature, applies generally to research on schizophrenia and particularly to the search for causal factors. If environmental factors turn out to be *interaction effects only*, the clues provided so far will continue to frustrate our search for universal conclusions. In other words, a factor may cause schizophrenia only in the relatively few individuals who are genetically predisposed to it, and such a factor therefore may have little schizo-specific effect in the general population. The histories of our schizophrenic twin samples generate a shopping list of precipitants. One twin was actually being trained to "expand his consciousness" after being recruited to a pseudoreligious cult; this turned out to be as disastrous as looking for a gas leak with a lighted match (and may well be generalizable to other vulnerable people in other cults). Childbirth itself was a clear precipitant in one twin's life: a schizophrenic break followed each delivery. Yet, in another pair, only the *childless* twin was a schizophrenic, and in another pair, the twin became psychotic after her second child but her sister, despite four pregnancies (including an abortion on psychiatric grounds), was never more dysfunctional than neurotic.

The great majority of us have endured the universal "trauma" of a vaginal delivery birth, but only 1 percent of us grow up to develop the syndrome of schizophrenia. One of our fraternal female pairs included a schizophrenic index case who was a breech delivery weighing only 3-$\frac{1}{2}$ pounds and with probable birth injuries leading to ptosis (droopy eyelids) and right leg impairment. Her sister, weighing in at 6-$\frac{3}{4}$ pounds, with no obstetrical complications, was attractive, successful, and a happy, mentally healthy mother at follow-up. Were it not for a hypothesized predisposition to developing schizophrenia, the index case — an ugly duckling if ever there was one — could have developed into an individual with any number of other forms of psychopathology or even into a handicapped but psychologically adjusted woman.

That this case vignette is not an example of circular thinking is supported by another twin history, this time from the offspring of the twins studied in Denmark by Gottesman and Bertelsen (see Chapter 6). Morten was quite normal and was engaged to be married when, at age 23, he fell 4 meters onto a stone cornice, was rendered unconscious for 5 minutes, and recovered with no direct consequences to his mental health. With time, he became more reserved and increasingly psychotic, with auditory hallucinations and persecutory and bizarre somatic delusions. He was admitted to a mental hospital on three occasions for a total of 43 years, always diagnosed as paranoid schizophrenic. His cotwin, Karl, married at the age of 19 and had three children. He was steadily employed until retirement age and, while prone to drinking, was otherwise normal until senile dementia developed at age 80, which progressed until his death at 88 from a pulmonary embolism. At no time during his 88 years were they any suggestions or symptoms of schizophrenia in his behavior. And now the interesting twist.

One of Karl's daughters, unmarried, developed an insidious psychosis at age 40 with incoherent speech, preoccupation with spiritualism, and social withdrawal. She lived with her demented father amid rubbish, decaying food, and human waste until her death from cancer at age 47. She was never admitted to a mental hospital, but was considered to be psychotic and probably schizophrenic based on a psychiatric interview in her home. In this scenario, the head injury to Morten appears to have been enough to release his predisposition to schizophrenia. His identical twin Karl never encountered a trigger for his diathesis, but he nonetheless transmitted it to his daughter, whose psychosis had no known precipitant. Without this kind of pedigree — discordant identical twins and their adult offspring — the schizophrenia seen in Morten after his head injury might have passed as one more instance of a *nongenetic phenocopy* of schizophrenia caused by an insult to the central nervous system (see Chapter 2). Now we have evidence both for brain injury as a releaser of a genotype for schizophrenia and for it generating schizophreniform psychoses in nonpredisposed persons who are merely phenocopies. Yes, "Seek simplicity and distrust it."

Summary

When human geneticists estimate the magnitude of genetic factors contributing to the liability for developing schizophrenia (see Figure 8), they obtain a value of about 70 percent for that statistic termed *heritability*. The importance of environmental factors in the liability to developing schizophrenia is *not* communicated by such a number. One of the major reasons that high heritability values do an injustice to the power of environmental contributors is that heritability is a concept adapted from agricultural population genetics, where the convention tells us that (100 − 70 percent), or 30 percent, indicates the strength of environmental variation in the population as a whole.

Even though the facts about a population of schizophrenics may be accurate enough, environmental factors of the sorts examined in this and the preceding chapter will be critical factors when they are experienced by an individual who is in the neighborhood of the threshold value for liability shown in Figure 8. Therefore, even individuals known to be at high risk for developing schizophrenia — such as the cotwins, siblings, or offspring of patients — could, in principle, avoid crossing the threshold into overt psychosis by avoiding, where possible, the kinds of environmental factors, both psychosocial and "ecological" (amphetamine, influenza, PCP, crack, EE, head injury, etc.) that add liability units to their schizophrenia balance sheet. By the same reasoning, those already over the threshold may return to the other side by reducing their liability units and increasing their assets.

Those few individuals in the general population or among the relatives of severely ill cases who happen to occupy a location at the extreme right tail of the distribution of *genetic* liability may easily become affected with schizophrenia or schizophrenia spectrum disorders. It is primarily such unlucky individuals who will not have objective, easily defined stressors prior to their breakdowns; for these unlucky people, almost any untoward event or mild brain insult can provide sufficient stress to trigger an episode of schizophrenia. However, for the vast majority of persons who are not so extremely disadvantaged genetically, the psychosocial and ecological stressors discussed above and in the previous chapters will play a major role in precipitating a case or, by removal, in precipitating a recovery. It is worth repeating that more than 95 percent of the population does not develop either schizophrenia or psychotic affective disorders even when they are exposed to severe life-threatening or ego-shattering experiences. Only about 1 percent of the population, those in the zone of combined schizophrenia liability near the threshold for overt schizophrenia, will experience a breakdown by the augmentation of their liabilities from psychosocial and environmental stressors.

The new, but rational, dogma that flows from the accumulation of evidence and theory thus far is that while the genes (in a "schizophrenia system-complex") are necessary for causing schizophrenia, they are not sufficient or adequate by themselves, *and* one or more environmental contributors are also necessary for schizophrenia, but they are not specific to it. From the research available so far, partially reviewed in this chapter, the clusters of significant environmental contributors to schizophrenia liability can be conceptualized as (1) insults to the brain, (2) demoralizing or threatening physical environments, (3) emotionally intrusive experiences, (4) emotionally demanding experiences, (5) affective and emotional understimulation ("institutionalism"), and (6) disruptions to attention and information processing. These suggested clusters can serve as guides to prevention and rehabilitation and to the avoidance of what Richard Day at the World Health Organization has termed "toxic environments."

9

Anguished Voices II
Personal Accounts of Family Members

The psychosocial contributions of the family and the surrounding physical and cultural milieu to the development of schizophrenia as detailed in the previous chapter were, for the most part, derived from the study of groups rather than of individuals. To recover the *individuals and their families* from the group data, we present a continuation of the personal accounts given in Chapter 3, but this time written by members of the families of a person suffering from schizophrenia. Such histories help to restore an integrated view of the processes involved in the dynamic interplay among patient, family, and the flux of life events. By complementing the drier, reductionistic approach taken by research workers, we hope to spotlight the complexities confronting the search for causes and ameliorations.

Making the Best of It
A former schoolteacher discovers the power of behavioral techniques in managing her chronically ill son at home and still manages courageously, at 65, to make a life for herself and her husband.[1]

When Dan first began to suffer from schizophrenia, our family thought it was just a case of teenage blues. We sent him off to college. By the time he began attacking the refrigerator for reading his mind and threatening family members for using the word "right," we had learned to recognize the disease. We read books. We attended lectures. We joined groups. But the first thing we did was go to our family doctor for advice. He was an honest and wise man. He told us there is no cure for schizophrenia, and the most important thing for us to remember was not to let Dan's problem destroy the family. We have six children. The two youngest at that time were 6 and 10 years old. Three are older than Dan. All our children were creative and full of fun. Christmas never went by without family skits and games. Summers we lived in the woods on the water, studying wild life, writing stories, planting gardens. It was a happy family, and I wish I could say it hasn't changed. Our family isn't destroyed, but it is badly damaged.

We did everything we could to be helpful to Dan. We signed him up for a specialist in New York, but after we had waited many weeks for the appointment to come due, Dan disappeared the day before we were to leave. We took him weekly to a psychiatrist, and then tried to see that he took his medicine. Whenever he was released from a hospital, we helped him find a job. We took him around until he found a suitable room in the area, and then once more helped him move his drafting table and other belongings to his new home.

He never took his medicine once he was away from us. Gradually he began working less and coming more often for "visits." Each time he finally began breaking things, and we once more brought the drafting table, suitcases, and sometimes cockroaches back home. Eventually, he would be reaccepted by a hospital, and we sat with heavy hearts while heaving sighs of relief.

---

[1]B. P. Piercy, *Schizophrenia Bulletin*, 1985, 11: 155–157.

Needless to say, we often neglected the younger children, and sometimes they were abused by their sick brother. I won't go into that now. It is water under the bridge. The other children are all married and living with their own families. My husband, John, and I have given up the hope of "golden years" together, though he occasionally says he wants to help Dan find a place before he dies. He is now 71 and has had two heart attacks. I am 65, in excellent health except for regular unexplainable seizures. We have had a few brief periods of togetherness without responsibility for our children, though there has never been a time when we would be surprised to look out the window and see Dan coming up the driveway. This was true especially in the year we were testing adult homes, though we continued calling on Dan, taking him places, and inviting him home for visits. Here is an example of Dan's life during this period:

On August 6, 198[ ], we drove Dan to Richmond, stopping on the way to buy him a new pair of shoes. Ten days later, John suggested we visit Dan for the day — take him out for dinner and go to a museum. We called to find out Dan was in jail. He had eaten a restaurant meal he couldn't pay for and assaulted a policeman.

The manager of the adult home explained Dan's problem to the police, and they released him, but Dan was not happy to be back at the home. He thought the food was terrible. Dan is a proud man. He wanted to have only the best. He would do anything to eat in restaurants. He sold his new shoes, pawned his watch, ran up a bill in a friendly Vietnamese cafe, and found new places where he could order a meal only to discover later that he didn't have his billfold. He was constantly asking John to raise his allowance from $20 to $35 a week. Whenever we went to visit him, he would spoil our time together by badgering his father. . . .

The following week, we went to Richmond to take Dan to the County Fair. He wasn't at the home. We looked in several eating places, went to the library, searching the streets on the way, and finally went to the Fair without him. When we got home, he was here. He had been on the way overnight and was starved. This time we let him stay home a week before taking him back.

In the next few weeks, Dan's trips home became more and more frequent. As Christmas approached, two of our children called to say they would not come to visit over Christmas holidays unless we could assure them that Dan would not be here. After all, they did have the joys of young children to consider. John felt very strongly about having a place for Dan to be for Christmas and said that he would be invited even if no one else came. Our two younger children wanted to be home. Here is the page from my diary for that December 25:

Dan was stomping, slamming doors, screaming until 4:00 A.M. Elizabeth and HC (her man) were nervous and sleepless and stayed in bed until 9:00. I was up early, found a burned skillet in the sink, a broken one in the trash. Oysters were on the wall and floor. I washed the dishes and prepared the turkey. Gradually people began to get up. We opened our presents, and then had a nice breakfast together. Elizabeth and HC left at noon. Fred went with his friends. At 6:00 P.M. Dan, John, and I had the 17-pound turkey and fixings. They were through eating in 10 minutes. After dishes, we watched the TV news.

Within the next few days, Dan was getting more and more upset over people "messing with his mind." He finally threatened to kill his younger brother who moved out to live temporarily with a friend.

It wasn't long before Dan was back in the hospital. Though the doctor said there was nothing wrong with him, his lawyer had talked him into volunteering.

While Dan was in the hospital, we once more investigated adult homes, hoping to find one which would please him. By the time he was to be dismissed, we hadn't found anything suitable, and in the end, we presented him with a plan for living at home. He now was 34 years old, and younger brother Fred was away at college, so it would be just the three of us. Here was our proposition:

We welcome you to live at home with us if you take your medicine regularly, eat at meal time, and smoke only in a restricted area. We will give you $30 a week for your expenses, and you will have occasional use of the car after you get your driver's license.

For over a year now, he has been going to the clinic to get his shots. He knows a good thing when he sees it. That $30 never makes it to the end of the week, but if he talks hard enough, he can always get a couple more dollars out of his Dad, who would rather toss him the bills than risk building up his blood pressure.

Three months ago, John and I began to wonder if we would be able to make our yearly trip south so that he wouldn't be housebound through the worst of the winter. Dan kept telling us that we should go and leave him here. I think I would never have done it, but John was so eager to go that he finally bought Dan a secondhand car, gave him the promise of $80 a week, and we headed for Florida.

It's hard to say who enjoyed the vacation most. It wasn't cheap, but luckily we were staying in the cabin of some friends, so all our extra expenses were up at this end. Dan curtailed his loneliness by making long distance calls all over the country. With his own car, he was able to keep his clinic appointments. He ate to please himself and kept his own sleeping hours. He was inspired to write several pages on his book about mental hospitals. Our allowance didn't quite cover his bills, of course, and he had to forge one check for $80, but he survived beautifully. In fact, it has been a little hard for him to adjust to our return home, but we've raised his allowance to cover the cost of his car, and he admits it's nice to eat my homemade soup. With summer coming on, the smoking problem will be alleviated. And now we all have a dream to look forward to next winter when Dan will once more be king of his castle, and John and I will share 2 blissful months alone together.

A Father's Thoughts
The informed views of a father who read what is said about schizophrenia and its treatment after coping with 11 episodes over the 9 years since his son became schizophrenic at age 19.[2]

My son Jim has been a paranoid schizophrenic for 9 years. Being retired, I have had time to study the literature about the illness in an attempt to understand it. It is an ambiguous project because each schizophrenic is different and there are many conflicting and overlapping theories. Many endogenous and exogenous factors seem to be involved in the development of schizophrenia, but I believe the evidence points to an endogenous imbalance within the central nervous system. Without such a necessary precondition, I do not believe a person would become schizophrenic whatever his life situation might be.

If so, why Jim should be ill is uncertain. There is no known history of the illness in either parental family, and there are two normal married sisters. The absence of a family history of schizophrenia probably does not rule out a biogenetic factor. Just as certain physically visible "birth defects" can occur in the absence of prior history, possibly an invisible "birth defect" can occur in the central nervous system.

Other factors may be involved in Jim's case. As opposed to his sisters, there was intrauterine complication, arduous birth, postbirth breathing block, and extreme colickiness. From what I have read, it is possible that these factors might be implicated in some way.

---

[2]Anonymous, *Schizophrenia Bulletin*, 1983, 9: 439–442.

Professionals have written of parents remembering their schizophrenic offspring as being normal in childhood, but usually with some questions as to parental retrospection or defensive reaction. I do remember Jim as a happy and alert child, and family relationships as being relatively good during his childhood. While I recognize that family life has emotional and attitudinal effects on family members, I do not believe our family life caused Jim's illness. No family is perfect, and it may be that some aspects of family life affect a preschizophrenic in a way that they would not if a biological vulnerability did not exist. I think that before Jim became overtly disturbed, we may have intuitively sensed that something was not quite right, and he in turn may have been negatively affected by our concern.

Possibly a second grade report card has some bearing. It stated as follows: Is not working up to ability, does not enter freely into class discussion, does not distinguish between work and play, and has difficulty with attention span. I have wondered if these characteristics, which are still apparent in adult life, might have been early behavioral markers of a predisposition for poor life adjustment.

Jim showed improvement during grade school because of his mother's coaching and encouragement, but in high school things worsened. While Jim made a fair start the first year, his grades deteriorated the second year and he ultimately was graduated with barely passing grades. I attribute his poor grades not to lack of intelligence but to innate attentional difficulty. Because his sisters and peers received average to excellent grades, I think his scholastic difficulty was a gross blow to his self-image and the start of serious self-doubt. I had several conferences at the high school; they said Jim was having a difficult adjustment but would grow out of it. Psychiatric treatment was not suggested. We accepted the school's assessment, but in retrospect, Jim's scholastic problems were warning signs.

After high school, Jim obtained a menial job and moved to an apartment with several other boys. Although he had been very much against drugs during high school, he experimented with "uppers" and "downers." After a year, Jim came home and quit taking drugs. His first hospitalization was over a year later. My wife and his sisters say the drugs "did it to him." I would like to believe this but feel Jim's developing illness drove him to experiment with drugs. The drugs may have had an effect on him, but I do not believe they were primarily responsible for his illness.

Something must be said about interpersonal relationships. Basically, Jim was a friendly person. He was not a "loner" and was motivated to be with people. He never wanted to upset anybody and did not easily tolerate in himself or others negative affect such as anger, disapproval, or criticism and even positive affect such as joviality. Jim had friends and a girlfriend during high school, but over the years, he seemed to drift to less desirable associates with whom he probably felt less stressed.

Just after high school, Jim told me he had trouble with people but then would not discuss his problem. He did not mean that he disliked people or was being treated badly. I think he felt unable to cope and sensed his emotional arousal as threatening to himself. At that time, Jim blamed himself for his problems and was not alienated from society. As pressures mounted, I think, he developed a great fear of himself. Later, self-blame and fear were rationalized and projected on to family and society in a paranoid way, leading finally to psychosis.

Since his first hospitalization at age 19, Jim has had 11 hospitalizations in 9 years. There was an 11-month stay at a state hospital during the sixth year. Two hospitalizations were by police action and three were by court order. Six times Jim went on his own to the local receiving hospital and was admitted. Jim must have felt an intense feeling of arousal that frightened him and prompted him to seek hospitalization. Once in the hospital, though, he would not open up to psychiatrists or social workers. Once he did mention fear of loss of self-control.

I think that earlier than we realized Jim had an intuition that something was not quite right. Jim told a favored cousin that he thought there was something wrong with himself, but we could not get him to seek help. He had a card on his desk that read, "If you need help, ask for it. If you don't, prove it." Later, I made another effort to get Jim to accept help, but it was rejected with hostile agitation. His mind was fighting for survival and the mind will not easily admit to a mental problem. Of course, after becoming psychotic, Jim did not think anything was wrong.

During the first 6 years of Jim's illness, we were unable to establish continuity of treatment or medication because of our lack of knowledge, turnover of psychiatrists at the local receiving hospital, Jim's actions, and other factors. He had two private psychiatrists during this period. There are doctors and doctors, and we had to learn this by harsh experience.

One mental health center psychiatrist took Jim "cold turkey" off the 200 mg [of Thorazine — chlorpromazine] that had been prescribed by his previous doctor. Jim had been doing pretty well on Thorazine but said it made him tired. While I had been told there would be no withdrawal effects, Jim experienced extreme hyperactivity for 3 sleepless days and nights followed by a 36-hour sleep. After his long sleep, Jim emerged a fairly alert and cheerful person. After several days, he became withdrawn and stuporous. A month later, after a minor operation requiring anesthesia, Jim became psychotic on the general ward. Transfer to the psychiatric ward and remedication took place, but this break marked a turning point for the worse and 2 years later Jim was sent to a state hospital.

State hospitals have unpleasant connotations, but I was pleasantly surprised at the facilities and treatment Jim received. After a period on the receiving ward, he was placed in an open cottage situation. Later he was moved to a "return to the community" program. Although this state hospital had long-term locked wards and a crisis ward, it was still an open place. The staff was very cooperative with the parents.

Jim's progress in the community return program was only fair, but he was placed in a halfway house in our home town. This move put Jim in good spirits, nonhostile, and nondelusionary. He called to tell us of the move and sounded like his old self. This positive change in mood and thought might illustrate a psychological placebo factor. Unfortunately, Jim slipped back into his usual "in-betweenness."

The transitional home was staffed almost entirely by female professionals, some of whom I respect greatly. There was one staff member I could not agree with. She told Jim that if he acted crazy, people would think he was crazy. It may be that at a certain stage this approach is workable with some patients. But I don't think a delusional person is conscious of the fact that he is having delusions. Delusions may be deliberate at an unconscious level, but the person does not consciously realize this during an active delusional period.

The home has an "on call" staff psychiatrist, but some patients are under the care of private psychiatrists. There has been antagonism between the home and some private psychiatrists, but the situation is easing now. I don't want to dwell on this friction except to say that some local psychiatrists are not as open-minded or progressive as they should be. In turn the home staff is suspicious of overmedication and is more therapy-minded. I think it is probable also that some patients are undermedicated.

Jim again did not respond well to group therapy. Maybe his lack of response can be related to his school history of not participating in class discussion. He did well in his work duties and was placed in an affiliated work program. He was also moved to a satellite boarding home. I was never sure whether he was moved because he needed less supervision or because of his negative attitude toward group therapy. It is a fact, though, that Jim has improved much more than some of the group-therapy-minded patients.

In the work program, Jim does janitorial work 4 hours a day in the community under the supervision of nonprofessional but trained work leaders. The work program has been good for Jim, but he did not like the boarding home and neither did we. After a year he asked if he could come home to live. Against professional advice, we allowed Jim to return home on the condition that he stay in the work program. Parents must sometimes be guided by their own feelings. I'm not sure we really wanted Jim back home, but we felt that rejection would have had a bad effect on Jim and on us. This arrangement has worked out fairly well. But living at home is a second choice. Jim would rather have his own apartment but can't for financial reasons. It is also questionable if he is ready to live by himself.

Considering the nebulous nature of mental illness, I believe the transitional home does a reasonable job. Jim benefited by being there. I was disturbed when the local newspaper published a letter to the editor from a private psychiatrist who was condemnatory of transitional homes. He seem to feel that if he could not successfully treat a patient in his office, the patient should be in a hospital.

After moving back home, Jim had a relapse requiring a 15-day stay on the psychiatric ward of a general hospital. This was in reaction to being reevaluated for Social Security disability pay. The threat of its loss and phone calls to him by the State representative apparently were too much for him. Perhaps this response illustrates his intolerance for stressful pressure. He was recertified and is back home and in his work program.

As a result of his last hospitalization, Jim came under the care of his present psychiatrist. He was released from the hospital taking Moban [molidone] and became very hyperactive mentally and physically. A lithium trial had no success. Both medications were stopped and he was put on Haldol [haloperidol], 30 mg per day. In 3 days there was remarkable improvement. In fact, there was some improvement after 1 day. The dosage was reduced after a year to 20 mg. Jim is doing well on Haldol and says it does not bother him as other medications did.

Jim's present psychiatrist seems more progressive than others he has had. The psychiatrist talks to Jim more and is more cooperative with rehabilitation programs. He has expanded his office to include a clinical psychologist and a social counselor. He has even given us his home phone number!

I believe the psychiatrist has been able to make Jim understand his need for medication. Jim is pretty good about his medication but rebels at too many pills or a difficult schedule. Right now he takes his medication at bedtime. I believe the medication is antipsychotic rather than antischizophrenic. The medication seems to reduce arousal, thereby relieving pressure on thinking functions. An adequate single dosage has always seemed to work best with Jim and Haldol seems to have the most favorable results with fewer side effects.

I have read about EE, the expressed emotion factor [this is the theoretical construct concerning negative or hostile feelings directed at the ill person, discussed in the previous chapter]. No doubt schizophrenics cannot easily handle expressions of concern, anger, disagreement, or disapproval — particularly from parents. Our experience with Jim has been that when he is disturbed by events outside the home, he becomes edgy and seeks assurance from us. It is sometimes difficult to find a middle course that does not support his fear and negative thought and still have him feel we are on his side.

Parents cannot go through years of living with the illness without feeling concern and disappointment. It's not really possible to hide this response. It is intuitively felt by Jim. I once said something to Jim with a smile, and he told me that a smile was no good if it was deceptive. On another occasion, Jim reacted to a smile by asking why I was laughing at him.

Delusions have been a problem. We have had mixed advice about how to deal with delusions and have had to learn by experience. Once when Jim was in a good mood, I tried to reason with him about a delusion. He said, "Dad, don't tell me. I know I started the EPA [Environmental Protection Agency]." This was said in all sincerity and good will. At other times hostility or withdrawal was the reaction.

Several weeks ago, Jim said he thought he should go into the hospital as he was having hateful thoughts. I told Jim that since he now realized he has such thoughts, he can disregard them. Jim told me not to counsel him as I was not a doctor. I then told Jim to call his doctor, which he did, and he did not go to the hospital.

I think UEE, unexpressed emotion, must also be considered. Despite his illness and resentment, Jim still trusts us more than anybody. In giving support to Jim, we have to be realistic even though disagreement may cause resentment, confusion, or withdrawal on his part. We have had to learn when to stand firm and when to compromise. Parents must guard against counter-withdrawal. This can cause resentment, and resentment in a schizophrenic can quickly turn to hostility. Jim may want to talk but is "frozen up." Saying something may "unfreeze" him or not. I have said something to him with no immediate reply and gone back to reading only to have him answer 5 minutes later. Silence can be deadening, and for Jim, probably results in too much brooding. Sometimes a mild display of temper is a steadying cue to him.

I have cited these problems to show that even when parents have an objective orientation to the illness, living with the ill person can be nerve-wrecking and exhausting. I am sure having an ill son has had a negative effect on us. We are probably somewhat affectively blunted now ourselves. But we survive: my wife is back teaching school and I keep busy with two hobbies, studying and trying to write.

Jim is now in his best state of remission ever, but will it last and be improved upon? There is no sure way of telling. As parents, we are frustrated because we are limited in how much we can be of help. So much depends on Jim's own mind. We have concern about his future, particularly after we are gone. We try to guard against Jim's becoming too dependent on us. We can only do the best we can while we are still here. Jim is quite rational now but very moody. He may for the first time be trying to adjust realistically to the conditions of his illness.

Jim mentions his inability to read or keep track of things on TV. But he drives a car very well. Apparently the type of concentration of attention required for driving is not the same as for other types of activity. He listens to music a lot. Music is not stressful to him. I like to play video games on occasion when he and I go out for coffee. He will not play them. One time he said they drive him crazy. I believe the concentration required stresses him. Later, he said he won't play video games because they waste too much electrical energy. This is not an illogical statement, per se, but is it a true belief or a rationalization to cover up his difficulty? I believe the latter and wonder if chronic defensive over-rationalization has something to do with causing thought disorder.

While families may be defensive about family history, one must be cautious about what schizophrenics say. Jim once told me that the only people who had ever treated him decently were his sisters, but later when he was hospitalized he said he was having problems because they beat him with frying pans. At other times, it was something his mother or I did. Schizophrenics look for reasons to justify their condition.

Many schizophrenics show resentment toward parents. Some amount of normal resentment develops during the strivings for independence of adolescence and is outgrown later. I observed this in my daughters. We never really restricted independence, but I think normal resentment is magnified in the schizophrenic mind.

I am not positive of my hypothesis in this essay. I see much similarity between myself and Jim, but whatever my faults, I have always had good attentional ability and tolerance for stress. This reinforces my belief that a biological deficiency is the root cause of Jim's illness.

Confessions of a Schizophrenic's Daughter

A teacher at Bellevue Psychiatric Hospital describes the chaotic life
she has led with a gifted and colorful schizophrenic mother now free
to wander the streets of New York as a "bag lady."[3]

My mother is a paranoid schizophrenic. In the past I was afraid to admit
it, but now that I've put it down on paper, I'll be able to say it again and
again. Mother, schizophrenic, Mother, paranoid, shame, guilt, Mother,
crazy, different, Mother, schizophrenia.

I have been teaching inpatient children on the children's ward of
Bellevue Psychiatric Hospital in New York City for 13 years, and yet I'm
still wary of revealing the nature of my mother's illness. When I tell my
friends about my mother, even psychiatrist friends, I regret my openness
and worry that they will find me peculiar.

My profession is appropriate for the daughter of a schizophrenic; at
least psychiatrists will think so. Since I often marveled that I escaped
being a disturbed child, I decided to devote my life to helping difficult
children. I have been successful in my work, which includes forming
relationships with the mothers of my students, especially the schizophrenic ones, whom I visit on the wards during their periods of hospitalization.

I was born in Kansas City, Missouri, in 1933. When I was 5 years old,
we moved to the Country Club section of the city, an area as spotlessly
bourgeois as any residential area in the United States. The inhabitants of
this region composed a homogeneous population of upper middle class
citizens, all very similar in their life styles. Not even one unusual person
could be found loitering on the streets of this hamlet, let alone a paranoid
schizophrenic. If, according to the laws of probability, there were schizophrenics and other "crazies" scattered about in the population, they were
well hidden.

On the outside our house resembled those of our neighbors, but on the
inside it was so different that there was no basis of comparison. Our
house was a disaster. Everything was a mess. Nothing matched, furniture
was broken, dishes were cracked, and there were coffee rings and cigarette burns clear across our grand piano. I was ashamed of our house. It

[3]R. Lanquetot, *Schizophrenia Bulletin*, 1984, 10: 467–471.

was impossible to bring friends home. I never knew what my mother might be doing or how she would look. She was totally unpredictable. At best she was working on a sculpture or practicing the piano, chain smoking and sipping stale coffee, with a dress too ragged to give to charity hanging from her emaciated body. At worst she was screaming at my father, still wearing her nightgown at 6 o'clock in the evening, a wild look on her face. I was never popular as a youngster, and I blamed my lack of popularity on my mother. . . .

Mother was quite interested in music and ballet, and she took me to every ballet and concert in Kansas City. She always looked terrible when she went out, and more than once she arrived at the theatre in her bedroom slippers. I was embarrassed to be seen with her, and before we left home, I would try to convince her to dress properly. She never listened and sometimes became angry, but chic or not, I accompanied her. I loved music and dance as much as she did. I even gave up Saturday afternoons to stay home with her and listen to the Metropolitan Opera broadcasts, and I loved her most and felt closest to her sitting in front of a gas fire, feeling her bony arm around my shoulders as we listened to the music together. Throughout my childhood I was torn between my bizarre, but loving, artist-mother and the conventional mothers of my friends.

Although Mother was rarely at home during the day, she could be found at the ballet studio. I think that I was probably born at the studio, because I can't imagine that Mother could have gotten to the hospital in time to deliver. Although she continued to take classes until her psychotic break, as soon as I was born she unconsciously decided that I should become the danseuse étoile that had been her goal in life. I didn't have the talent to be promoted to such heights, but failing to understand this, she continued to nag me to take more classes and work harder.

Feelings of shame and fear overwhelmed me in those early years, shame that my friends would find out that my mother was "different" and fear that I would be "different" too. The fear of being like Mother must have prevented me from studying ballet and piano seriously. My Mother played the piano and danced, and she was schizophrenic. If I played the piano and danced, I would be schizophrenic also. I was terrified that if I showed any signs of letting myself go and really working, my mother would close the doors of the studio and fasten them with a heavy, iron bar. . . .

There were other problems in living with a schizophrenic mother. One was the lack of tranquility at home, the commotion, and chaos. My parents were constantly arguing about money. My mother had no idea of budgeting. She didn't need to learn, because her father was available to supply her with money as needed. My father didn't approve of limitless concert-going, dance classes, or book buying. He abhorred eating in restaurants everyday and expected Mother to stay home to take care of the house and prepare supper. I would be awakened at night by screaming and lie in bed pretending to be asleep, morbidly fascinated by my parents' quarrels. . . .

One day, when I was 10, my mother vanished, and as if by magic, my father moved back to the house to take care of us. I resented his return. He had abandoned us, and I must have felt that he was responsible for Mother's problems. We were told that Mother was ill in a hospital in Burbank, California, where my grandfather's sister was a staff physician. I felt very lonely without her and began hanging around the ballet studio. Once the teacher put her arm around me and said, "Poor child, you miss your mother, don't you?"

Years later I learned that Mother had run away to New York without telling anyone she was leaving. She was making frenzied visits to the ballet schools there when a friend of the family phoned my grandparents to inform them of their daughter's strange behavior. My grandparents immediately set out for New York to rescue Mother. They brought her to Menninger's Clinic, which had not been in existence very long. At the time the hospital was located in old-fashioned red brick buildings that were already on the premises when the Menningers moved in. Equating building height and glass walls with hospital excellence, my grandparents took one look at the hospital and headed for California, where Mother was hospitalized for a year. She regained her physical health, but her mental health was totally ignored. When she was discharged, we joined her in California, where we lived for the next 2 years. Mother was subdued and withdrawn from any human contacts outside of the family. Her withdrawal was less of a bane to our social life than her neurotic existence in Kansas City, but she lost something of the artist, her most interesting self. . . .

At the end of my junior year in high school I was in a serious automobile accident. I tend to think that my mother's consequent decompensation might have been precipitated by my being in a coma for 6 days, but I'm not certain. After I came home from the hospital, she became very strict with me, although she had never interfered in my social life previously. When I protested against her arbitrary, nonsensical restrictions on my dating, we began to have terrible fights. I could not make her accept the fact that a monastic existence was not for me.

Mother and I shared a room with twin beds. When Mother was lying down, she would start to moan as if she were talking in her sleep. "I can't stand that girl. She's evil; she's a bitch. She's just like her father." I was terrorized, but I dared not move. I felt I had to pretend to be asleep, because I didn't want her to know I was listening. I tried to deny the reality of Mother's illness by not acknowledging the outbursts. I used to lie in bed, wishing I were dead, believing that I was the worthless girl she was describing. . . .

I still remember with horror the night that I came in late from a date and decided to sleep on the couch in the living room in order not to wake Mother. I made an effort to avoid disturbing her, not because I was being considerate, but because I didn't want her to start moaning. As soon as I lay down, she came into the room and stood next to me, calling me a prostitute. When she spat on me, I grabbed her upper arm and bit it as hard as I could. The outline of my teeth etched in black and blue remained visible for over a week, but Mother never mentioned it. Even now when I think of the incident, I feel shame because of my loss of self-control and display of aggression toward my poor, defenseless, crazy mother.

Next Mother began to insult strangers on the street. She would stop in front of a well-dressed bourgeois of Kansas City, fix her eyes on him for a few seconds, and snap angrily. "What's wrong with you? Why are you looking at me like that. I'm going to tell my lawyer." If my brother or I were with her, we'd be so embarrassed that we'd want to disappear into a crack in the sidewalk. No matter what we did, she wouldn't stop. Once she hit someone over the head with her pocketbook and another time notified the police that the neighbors were spying on her although they'd been gone for 3 months. In the sterile atmosphere of Kansas City, her outbursts upset everyone. In New York she wouldn't have been noticed.

My choice of colleges was based on their distance from Kansas City and Mother. I had to get away before I became crazy. I applied to the University of Chicago, Barnard, and Stanford and was accepted at all three. My grandfather refused to let me attend Chicago U. He said that Chicago was no place for a young girl, but I knew he refused because of Mother, who had attended the Chicago conservatory for 3 months before she returned home to Daddy. I decided against Barnard, because Mother liked New York. I was afraid that she might follow me there. That left Stanford.

Much to my dismay, Mother arrived in San Francisco during my sophomore year at Stanford. She came for a visit and decided to stay. The fantasy about being haunted by the specter of my schizophrenic mother had come true. She moved into a dumpy apartment two blocks from a dance studio and began taking Flamenco dancing from a Spaniard who taught there. She fell in love with her teacher, but he didn't care about her. Although he paid her less attention than he did the other students, she was always hanging around, gazing at him in abject adoration. She never realized how pathetic and absurd she appeared. . . .

Mother's descent into chronic schizophrenia would take too long to describe. She was finally admitted to Menninger's where she improved during the first year and a half of treatment. Then her father died. She had always been her father's little girl, and her universe was shattered without him. The only person in the world she trusted had departed. After the funeral she refused to return to Menninger's. She had learned that no one had the right to send her to an out-of-state hospital against her will.

Two years later she had to return to the hospital. She was driving a car without brakes and insulting black people by loudly declaiming her theories about the inferiority of the black race. By her second admission it was too late. Mother had become a chronic schizophrenic. After the first year of hospitalization we were asked to remove her. The Menningers were only interested in patients whom they could cure. The family, for we remained a close family, rallied its forces to find another hospital, and we transferred her to a Mennonite Hospital in a small town in Western Kansas. We were grieved by the loss of Mother as a functional human being, a bereavement that was finalized by attaching "chronic" to her diagnostic category. I was especially horrified by the necessity of burying my mother in the country. My trepidation was intensified when Mother was moved to a halfway house near the hospital, where the only activity available to patients was filling mattresses.

When Mother began to threaten the doctor at the new hospital to find a lawyer to sue him, the family was forced to make the difficult decision to go to court to have her declared "incompetent," to openly admit that she was psychotic. We had to safeguard her trust fund from a shyster lawyer. My grandmother was devastated. Having avoided the truth for the greater part of her daughter's life, she couldn't face the fact that Bonnie was crazy. Since I was in Europe, my brother, my uncle, and the doctor testified that Mother was a danger to herself, Mother was declared "incompetent." Mother lost all liberty, all sense of self. Any step she took had to be authorized by the magistrate.

Once drug therapy came into being, Mother was force-fed Haldol [haloperidol] against her wishes, and this resulted in a remission of symptoms. Feeling well enough to leave the hospital, she made the decision to go to Menninger's by herself to take an examination that she supposed would disprove her insanity. Of course, her guardian was forced by law to make her return to the hospital. She had not requested the judge's permission to make the journey. Later my brother accompanied her to Menninger's for an evaluation, the results of which showed that she was well enough to leave the halfway house in Kansas and come to New York to be near her children. They specified, however, that she would need to live in a structured environment.

Eventually the family received the authorization to have Mother's guardianship transferred to New York, and a "Committee" was appointed by the New York court. When Mother first joined us in the East, she behaved the way she did when she was discharged from the hospital in California—withdrawn, isolated from everyone but the family, yet able to profit from all the big city had to offer. Listening to music or watching ballet, she came back to life. Vital energy that had been absent for so long returned to her body. The results of changing her habitat were much better than we had ever expected.

Not only did Mother rediscover art and music in New York, but she soon became familiar with the liberal New York laws regarding "patients' rights." She refused to continue to take Haldol and slowly began the reverse trip to "No Man's Land," where she now dwells. The first sign of her decompensation was a refusal to come to my apartment, and then she rejected me completely. Next the manager of her middle-class apartment hotel asked us to remove her. She was annoying the guests with her outbursts. She had become known to all the shopkeepers on the block as "The Crazy Lady of West 72nd Street." Looking like a zombie, she paraded down West 72nd Street, accusing aunts, uncles, and brother of stealing her father's fortune, screaming at people who frightened her, discernible from her New York counterparts only by a Midwestern accent and an absence of curse words.

Having been told over and over again in our youth that it was our duty to take care of Mother, my brother and I initially resented our burden. We felt that since Mother had not accepted the responsibility of her children, we should not have to be responsible for her. At that time it was difficult to admit that we actually loved our frail, unbalanced mother and wanted to help her. When we grew up, we began to understand why Mother was different, and our resentment lessened. On Haldol Mother's behavior improved tremendously, and we even harbored false hopes of her return to normal living. We never suspected that she might cease taking medication and regress. Whether or not it's preferable for her to be forcefed Haldol and incarcerated in Kansas or allowed to do as she pleases in liberal New York, as destructive as her life is now, is paradoxical. She was not able to enjoy life and pursue her artistic interests in the former situation, but she is even less able to do so in the latter. Without medication, she can only exist. I believe that basically she is less free in her present life, a prisoner of her delusions and paranoia. My brother, however, disagrees. He thinks that Mother is better off having the choice to live as she wishes, wandering aimlessly in the streets, constructing the world to fit her delusions.

10

Schizophrenia, Society, and Social Policy

A more comprehensive appreciation and understanding of the syndrome of schizophrenia, with all the ramifications of society's impact on it, and of schizophrenia's impact on society, can be derived from a consideration of the facts about patterns of mortality (death) and morbidity (disease susceptibility), crime and violence, and aspects of reproduction (sexuality, fertility or "fitness," marriage, and divorce). Closely related topics such as the laws dealing with the marriage and reproduction of schizophrenics and the theory and practice of genetic counseling will also be broached in this chapter. All these facts fall under the umbrella of the definition of *social biology*—knowledge about the biological and sociocultural factors that influence the structure of human populations—and have direct implications for implementing social policy.

Society depends upon the maintenance of the public health for its very survival; hence the term *vital statistics* for the aggregation of data on epidemiology and demography. Social biology, informed by the epidemiology and the demography of schizophrenia, can be used in the service of public health promotion and specific schizophrenia prevention (primary prevention), in early recognition of at-risk individuals and consequent intervention (secondary prevention), and in limiting disability and promoting rehabilitation (tertiary prevention). Regrettably, social biology can be perverted to evil ends and become "political (pseudo) biology," as it was by Adolf Hitler and the Nazis in implementing their insane policies of murdering psychiatric patients, genocide, and the Holocaust (Kevles, 1985; Lifton, 1986; Müller-Hill, 1988; Proctor, 1988). Only a sampling of the more interesting and important topics can be provided in this chapter (see Ødegaard, 1975; Saugstad, 1989; Allebeck, 1989; Lewis, 1989; Taylor, 1987), but many of these basic facts will help to dispel some of the myths that have accumulated about the syndrome of schizophrenia and will help to promote the welfare of those persons suffering from the syndrome by informing social policy decisions.

Patterns of Death and Physical Illness

All psychiatric patients seen in hospitals during the nineteenth and twentieth centuries, including those with schizophrenia, had markedly higher death rates across all ages than did members of the general population. Life insurance companies would have lost money if they had specialized in insuring the lives of hospitalized psychiatric patients and had used the standard mortality tables for men and women in the general population as their reference base for calculating premiums. For the first 70 years of this century, the excess mortality (compared to the rate expected in the general population at the same time) of schizophrenics ranged from 200 to 500 percent. In Sweden, where national vital statistics have been collected since late in the eighteenth century, a male schizophrenic, once he was admitted to a psychiatric hospital during the first half of this century, would have a remaining life expectancy of 68 percent of that of a nonhospitalized male; for female schizophrenics, the figure was only 54 percent (Larsson and Sjögren, 1954).

Tuberculosis (TB) was often pointed to as the major cause of the excess mortality. It was well known as a killer among institutionalized populations of all varieties and accounted for one-third of all deaths of schizophrenics before World War II. In fact, the observation led to speculations about a necessary relationship between the genes for schizophrenia and the genes for susceptibility to TB (Kallmann, 1938). Throughout the Western world of modern industrialized societies, TB has disappeared as a major cause of mortality among schizophrenics, but as recently as 1955 it had led to an excess mortality from 7 to 13 times the rate among all psychotic patients. As early as 1942, Carl Alström, a public-health-minded Swedish psychiatrist, had concluded that the significant loss of weight in schizophrenics admitted to hospital lowered their resistance and increased their susceptibility to TB. One of the side effects of the phenothiazine drugs introduced in 1954 for the treatment of schizophrenic symptoms was weight gain; the drugs seem to have eliminated the scourge of TB among schizophrenics.

In a comprehensive survey entitled "Schizophrenia: A Life-Shortening Disease," Peter Allebeck, an expert in social medicine at the Karolinska Institute in Stockholm, reports that even today excess mortality among schizophrenics is twice that of the general population. Like Letten Saugstad and Ørnulv Ødegaard of the Norwegian National Psychiatric Case Register before him (1979), Allebeck was impressed with suicides, accidents, and cardiovascular disease as sources of excess mortality, with cancer as a controversial conundrum. In the analyses of the causes of death for 10,000 psychiatric patients residing in Norwegian mental hospitals in the years 1950–1974, it was clear that cardiovascular disease among schizophrenics led to an excess mortality rate of 50 percent in females and 80 percent in males and showed a pattern of increasing mortality. Allebeck was able to perform an in-depth analysis of the causes of death among all 1,190 schizophrenics discharged in Stockholm County and followed up for 10 years to 1981. Comparison data for normals were available through the various national registers for vital statistics, easily permitting calculation of standardized rates by age and sex. Cardiovascular disease accounted for 37 percent of the 231 observed deaths in the cohort and was associated with an excess mortality of 80 percent. It may be a paradox of treating schizophrenics with phenothiazines, which relieve their stress and hyperactivity and increase their sedentary habits and weight, that the price paid is an excess mortality from heart disease. Insufficient information is available about the nicotine addiction of psychiatric inpatients to know its effect upon the rates for heart disease and cancer; most clinicians, however, would report that, if anything, more patients than nonpatients smoke, and they smoke excessively.

Cancer

Data on deaths from cancer in schizophrenics are intriguing. No excess mortality was recorded for Norwegian schizophrenics between the study years of 1950 and 1974; if anything, there was a suggestion of *decreased mortality* (40 percent less than that of the general population). Allebeck found no significant difference in the recent Swedish sample, but noted that it would take a very large sample followed for many years to reach any strong conclusions about decreased or increased cancer mortality.

Such a study is now available from Denmark, where the various psychiatric and population registers have been matched against a national cancer register that dates back to 1943 and classifies tumors by site. An eligible cohort of more than 6,000 male and female schizophrenics identified in a 1957 census has been followed through 1980 for a total of 100,000 person-years of observation for causes of death. Annelise Dupont, former head of the Institute for Psychiatric Demography, and her colleagues reported a *decreased* rate of cancer mortality in both male and female schizophrenics, but only the male rate was reliable. The risk in males across all cancer sites was 67 percent of normal; for females, it was 92 percent. More exciting data emerged when the neoplasms (tumors) at specific sites were examined. The most striking reduction was seen for cancer of the lung in males; the rate was 38 percent that of the normal population. The rate is even more striking if the clinical impression that psychiatric inpatients smoke more than nonpatients can be documented. Digestive tract and prostate cancers were also significantly decreased in schizophrenic males. The patterns were less notable among the females, but even here the reduced risk of mortality from cancers of the respiratory tract (38 percent of normal) and of the cervix (59 percent of normal) was statistically reliable. The reduction in cervical cancer may be explainable by the decrease in sexual activity among female schizophrenics as a group.

As long ago as 1909, the Commissioners in Lunacy for England and Wales had noted that psychiatric patients appeared to be relatively immune to cancer, despite the relatively crude stage of knowledge about cancer. In the careful Danish work, we seem to have very reliable information on reduced cancer mortality, especially at some sites, in schizophrenics. What do these data tell us? Psychotropic drugs used to treat schizophrenics may be protecting them from developing cancers; work with laboratory mammals has shown antitumor activity for many such drugs. Clues to both the biological mechanisms underlying schizophrenia and tumor formation may result from such research programs; more than one process may well be involved, given the possible decrease in cancer mortality before the introduction of psychotropic drugs.

Suicide

Suicide is *the* most frequent and alarming cause of excess mortality among schizophrenics, far surpassing any other cause. Although received wisdom would correctly implicate suicide as a major cause of death among sufferers from the major affective disorders (depression and manic-depression), it comes as a surprise to learn that schizophrenics kill themselves almost as frequently. In the World Health Organization 10-country follow-up of 1,065 patients with various severe mental disorders, Norman Sartorius and colleagues found that schizophrenics constituted 80 percent of the fatalities in the first five years. Of the 19 suicides, 14 had a diagnosis of schizophrenia, indicating that the risk of suicide is as great as or greater than that among patients with affective disorders.

The Norwegian series of deaths *in hospital* showed violent death to be the major cause of excess mortality, with suicide leading the list. For male schizophrenics, the increased rate went from 120 percent to 360 percent of "normal" between 1950 and 1974. The detailed follow-up of the Stockholm County register by Allebeck reveals an excess mortality rate among male schizophrenics of 1,000 percent of normal and, for females, of 1,800 percent of normal. Careful scrutiny of the death certificates and of reports of suicidal thoughts preceding the event, together with the 10-year posthospital follow-up period, helped to attain accuracy. The female rate of excess mortality is so high because suicide in nonpsychiatrically ill women is quite rare compared to suicide in nonpsychiatrically ill men; something about the process of schizophrenia has evened or even reversed the usual sex ratio for suicide. Efforts to identify the suicides retrospectively proved futile; Allebeck concluded that "suicidal acts among schizophrenic patients often are impulsive and difficult to predict" (p. 87). Manfed Bleuler had earlier discovered that suicides among schizophrenics occur at any time in the course of the illness — both during remission and during acute episodes.

I know and fear the danger of suicide late in the course of schizophrenias. The older view corresponded to the concept that in the course of time the inner life of schizophrenics "was extinguished, dulled, or burned out," and that in time they would lose their "internal dynamics" and their capacity to suffer. But the fact that decades after onset of illness suicides continued to occur points to the fallacy of that outdated assumption. (1978, p. 106)

A recurring observation made by clinicians working with schizophrenics is that the symptoms of depression and anhedonia (absence of pleasurable feelings) often co-occur at such a high rate that their presence is not a useful clue to the prediction of suicide attempts. However, differentiating the "facets" of depression so that a particular feature — pervasive hopelessness — can be discerned may well provide a clinical clue worth attending to as a harbinger of suicidal behavior.

Rheumatoid Arthritis

According to J. A. Baldwin, a clinical epidemiologist who maintains the Oxford (England) record linkage system covering a total regional population of 800,000 persons, of whom 2,314 were known to be schizophrenic, at least 20 diseases have been claimed to occur more often than expected in schizophrenics, and a further 6 have been said to occur less often. His review of the literature, as well as his own growing data base, led him to conclude that

Most of the suggested antagonisms and supposed deficiencies in schizophrenia, including cancers, epilepsies, allergies, diabetes, and myasthenia gravis, have arisen from interpretations of clinical "non-experience" based on more or less impressionistic overestimations of whatever is to be expected by chance. There is not as yet [i.e., 1979] sufficient sound epidemiological evidence for any of them, and the only negative association for which evidence is now fairly strong is that with rheumatoid arthritis. (1979, p. 617).

Although Baldwin's conclusions may have to be modified in regard to cancer, based on the new Danish study, the alleged protection from rheumatoid arthritis has continued to appear in various reports. If confirmed in large samples diagnosed according to the more objective schemes now available, such a finding could provide important clues to genes directly associated with physiological processes not now recognized as important to schizophrenia. Alternatively or additionally, it could provide clues to genes linked on the same chromosome to genes involved in the etiology of rheumatoid arthritis, but not yet discovered. Reports have persisted since 1936 about the low incidence of rheumatoid arthritis in persons with schizophrenia; combining the schizophrenics in both the Oxford and the Stockholm registers above, only 3 cases of arthritis were observed when 12 would have been expected among 3,500 schizophrenics (Spector and Silman, 1987). Clues to etiology are welcome from any source, however unlikely. The puzzling negative association here may renew the interest in immunological-viral theories about schizophrenia, including attention to the major histocompatibility region of chromosome 6 (McGuffin and Sturt, 1986) and to the role of endogenous opioids (endorphins), which are decreased in arthritis but increased in schizophrenics. At any rate, the reasons that schizophrenics do not appear to develop rheumatoid arthritis may provide leads to the genetics and pathophysiology of schizophrenia (Vinogradov, Gottesman, and Moises, 1991).

Studies of special individuals can be as rewarding as studies of the entire national register of schizophrenics in a Scandinavian country.

Among the series of identical twin pairs discordant for schizophrenia at Saint Elizabeths Hospital in the research being conducted by Torrey, Gottesman, and colleagues (see Chapter 6) is an intriguing pair now discordant for 13 years. Only the proband suffered from asthma from age 10, and it was he who had a gradual onset of psychotic symptoms in college, eventually resulting in hospitalization for schizophrenia at age 31. His identical cotwin has been normal and remains that way today at age 44; however, only the cotwin was diagnosed, at the age of 7, as having rheumatoid arthritis. We have a mystery: same genes, same prenatal and postnatal rearing conditions, but discordance for and protection from schizophrenia in the twin with rheumatoid arthritis. Sherlock Holmes, where are you when we need you?

Criminality and Violence

Obtaining a balanced perspective about the relationship between violence and schizophrenia is extremely difficult. Pamela Taylor, a forensic psychiatrist at the Institute of Psychiatry in London, has commented: "There is no doubt that schizophrenics are capable of violent behavior and, there, any certainty about the relationship between schizophrenia and violence ends." (1982, p. 269). Media distortions lead to an exaggeration of any genuine relationships that may exist, but dispassionate sources of information are available to help us attain a more accurate perspective.

When John Hinckley, Jr., a diagnosed schizophrenic, almost succeeded in assassinating President Ronald Reagan in 1981, history was repeating itself. One hundred years earlier, in 1881, Charles Guiteau mortally shot President James Garfield in the belief that the president was a threat to the country and that killing him had become a "political necessity" and "an act of God." After the event, Guiteau asserted that, because he had been controlled by "divine pressure," he was not legally responsible. He fully expected to be acquitted and to run for the presidency himself in the next election. The jury saw it otherwise; he was found guilty and executed in 1882, still convinced that the assassination had saved the country.

The White House, the Vatican, and similar centers of immense power seem to act like a magnet, attracting individuals suffering from paranoid schizophrenia and delusional disorders (paranoia). About 100 persons each year approach the White House to demand relief from imagined persecution, to offer advice on how to run the country, to obtain money, and so forth and are arrested by the Secret Service and sent to Saint Elizabeths Hospital for evaluation. David Shore, a psychiatrist with the

NIMH Schizophrenia Research Branch, and his colleagues have reported on the diagnoses of the 328 "White House cases" seen over a three-year period in the 1970s — 91 percent were found to have schizophrenia (66 percent paranoid schizophrenia) or a paranoid state. Three earlier evaluations since 1943 with smaller samples had reported similar results. It is somewhat reassuring that none of these 328 actually shot a president (although one later murdered a Secret Service agent). Thirty-one of the 217 males had murder or assault arrests by the time of a 12-year follow-up in 1988 (including one assault on a woman erroneously perceived to be the First Lady).

Other indelible acts of violence come to mind simply by mentioning the serial killers "Son of Sam" and Charles Manson, the assassination of Robert Kennedy by Sirhan Sirhan, the murder of John Lennon of the Beatles by Mark David Chapman, and the mass murder of Stockton, California, schoolchildren in 1989 by a paranoid schizophrenic on Social Security disability who opened up on a school yard with an AK-47 assault rifle. Schizophrenia or delusional disorder is a likely or a confirmed diagnosis for each of these individuals, based on the information available to the general public in media accounts. Knowing that a person has some kind of mental disorder does *not* permit an accurate prediction about their dangerousness, however. John Monahan, psychologist and Professor of Law at the University of Virginia, a recognized expert witness on this topic, has concluded that of every three mentally disordered persons predicted to be violent by an expert psychologist or psychiatrist, only one will go on to commit a violent act. In his experience, psychiatric diagnosis or personality traits are the poorest predictors of violence. The best predictors include the same demographic factors found useful in predicting violence in the general population of the United States: age, sex, social class, and a prior history of violence.

Before reaching any strong conclusions about the likelihood of schizophrenia and crime and violence going hand in hand, it is necessary to examine the larger crime picture. In the United States, more than one and a half million violent crimes were known to the Federal Bureau of Investigation in 1986 (the latest data), including 20,600 murders, 90,400 rapes, 543,000 robberies, 834,000 aggravated assaults, and 87,600 acts of arson. Such figures, impressive as they are of the risk of bodily harm, are merely a hint of the true level of violence in the United States. From public reports of victimization surveys by the U.S. Bureau of Justice for 1985 alone, we learn that there were a total of almost *20 million violent and nonviolent crimes*, most of which were not reported to the police. The seriously mentally ill made a trivial contribution to such widespread mayhem; you are safer visiting a patient in a mental hospital than you are on the streets of any major American city after dark. From this vantage point, the life-threatening behavior of some schizophrenics can be seen as isolated episodes.

Cross-national comparisons of the violent behavior of those with schizophrenia do not permit generalizations, because of the vast differences in the patterns and frequencies of crime, but some studies are informative. Eve Johnstone and her colleagues, in their London-area study of the problems encountered in the first episode of schizophrenia leading to hospital admission, provided unbiased information about the disturbed behaviors of 253 schizophrenics who met the criteria used in the WHO studies (Chapter 2). Six percent had repeatedly threatened the lives of others, as had a further 13 percent on one or two occasions; none had committed murder, but one almost succeeded when he inflicted multiple knife wounds on his father, believing him to be the Devil. Altogether, 22 percent had had contact with the police for bizarre or inappropriate behavior as part of their first episode of schizophrenia, including such events as the strangulation of a pet canary, cutting off the heads of the flowers in the family garden, and a male graduate student walking through a quiet suburb in lady's underpants embracing each exposed garbage can. At least half the families of a different sample of schizophrenics in England reported one or another kind of moderately to severely distressing, threatening, or noisy behavior in the preceding month (Gibbons, 1984).

Given that a person is ill with schizophrenia, what is the chance that he or she will commit murder or manslaughter? The answer requires determining the total number of violent offenses and offenders, the total number of mentally disordered violent offenders and their diagnoses, and the total number of mentally disordered but nonviolent persons and their diagnoses. Fortunately, H. Häfner and W. Böker of the Department of Social Psychiatry at the University of Heidelberg in the Federal Republic of Germany have succeeded in collecting such necessary information over a 10-year period (1955–1964) in the greater Mannheim area. Schizophrenics accounted for 284 (53 percent) of the 533 mentally disordered violent offenders.

In their research, Häfner and Böker discovered that, given that a man is known to be a schizophrenic in the Federal Republic of Germany, the probability that he will commit or attempt homicide is 0.05 percent or only 5 in 10,000 schizophrenic men will become violent offenders. The same calculations lead to figures of 0.006 percent, or 6 in 100,000, for those with affective psychoses and the same probability for those with mental retardation. As noted earlier by Monahan, age (14–30) and being male exerted a greater influence on the risk of becoming a violent offender than did psychiatric diagnosis itself.

Important information for the families of the mentally ill also emerged from the German study. Only 16 percent of schizophrenics who became violent offenders committed their acts during the first year of illness; 25 percent became violent between 5 and 10 years after onset, while a further 25 percent did not become violent until even later. The victims of the schizophrenics' violence were most often their own parents, siblings, spouses, and children (58 percent); friends (22 percent); and, lastly, strangers or authority figures (19 percent). This contrasts sharply with the assaults of affectively psychotic persons, whose victims were relatives 95 percent of the time. It is sad to relate that the two most sensational crimes in this decade of study were committed by paranoid schizophrenics. A 43-year-old offender, motivated to get revenge for delusional wrongs from society, constructed a flamethrower and killed 10 children and 2 teachers in a school, inflicted major burns on another 22 people, and then committed suicide. The other case, a 22-year-old paranoid schizophrenic, hit 10 strangers on their heads with a hammer in response to commanding hallucinatory voices.

Careful research of the kind conducted in Germany needs to be done in the United States, with its markedly different pattern of violence, to determine the true relationship between mental illness and violence. Only then will society be able to strike a balance between protecting itself from the mentally disordered and protecting the mentally disordered from society. England and Scotland have maximum security hospitals in addition to traditional mental hospitals. Of the approximately 2,100 inmates, 60 percent have a diagnosis of schizophrenia or paranoid state and a further 6 percent carry a diagnosis of manic psychosis (Taylor, 1987). Unlike criminal violence in Western Europe, problems in the United States are impacted by poverty, undereducation, racial factors, easy access to weapons, and endemic urban drug trafficking. Questions about the importance of mental disorders in crime, and of crime in mental disorders, need to be resolved on a national basis, informed by cross-national data.

Sexuality, Marriage, Fertility, and Divorce

In one of the rare articles in the scientific literature about the sexual conduct of schizophrenics, J. P. McEvoy and colleagues report on the attitudes of chronically schizophrenic women in a Tennessee state hospital in the 1980s toward sex, pregnancy, and birth control. Some 80 percent of these women, aged 20 to 58, wanted an active sex life, and 65 percent said that they had had sexual intercourse while in the hospital during the preceding three months. It would be very important to society to determine how generalizable such findings may be, because vital policy decisions about patients' civil rights are guided by such facts. E. Fuller Torrey, a wise and seasoned psychiatrist, has noted that once pregnancy in a schizophrenic women has taken place, "the couple and their families are often caught between a rock and a hard place" and urges that abortion and adoption both be considered as options. Torrey also suggests that the burden of decision-making be shared among all the "helpers" (from lawyer to social worker to religious adviser) to alleviate the emotional burden on all those involved, including the patient and her family.

Clinical impressions about the sexual behavior of schizophrenic men and women are usually so idiosyncratically anecdotal that they provide no guarantees about their usefulness in understanding other cases. Manfred Bleuler did take the trouble to explore this sensitive area in his long-term follow-up studies of 208 (108 women and 100 men) Swiss schizophrenia patients. Granted that the sexual mores of Zurich inhabitants may differ from those of other locales, Bleuler categorized the prehospital erotic life of his patients into four categories:

Nonerotic: no love relationship and no intercourse — 24 men and 40 women

Erotically discreet: "natural love relationships" — 43 men and 14 women

Erotically active: multiple sexual relationships, sometimes producing illegitimate children — 16 men and 42 women

Sexually "perverted": voyeur, homosexual, and incest with brother —one of each

The remaining 26 patients could not be categorized from the information available, but Bleuler was quite impressed with the frequency of celibate schizophrenics and the rarity of "perverts." A false impression about the latter is easily generated by the attention paid to exceptions in the media.

Before the onset of their illness, 68 schizophrenics had married; 11 of 28 men reported happy and successful marriages before symptoms appeared, as did 13 of 40 women; the remainder, 65 percent, said they had poor marriages. Although this sample of 208 patients had produced 184 children, 80 percent were born *before* the first hospitalization for the psychosis. Only 15 children were illegitimate, thus showing more restraint on the part of schizophrenics than among the general population of Zurich, where the rate was higher. An overview of the marital statuses of the schizophrenics contrasted with that of the general Zurich population is given in Table 15. It is clear that there is something very special about the process of schizophrenia that leads to a low probability of successful marriage.

Both affectively disordered and schizophrenic women were studied in the greater London area by Barbara Stevens to cast more light on the sexuality of psychiatric patients. Although 20 percent of the 843 children produced by 813 schizophrenic women were born out of wedlock, the same was true for 15 percent of the children born to normal women in the London area at the time. Careful life histories taken from the patients suggested that no more than 3 percent could be described as promiscuous, while 15 percent were asexual and 2 percent were lesbians. Most of the out-of-wedlock children resulted from a serious relationship, but the symptoms of schizophrenia diminished the chances of marriage before admission to the hospital. Fully half of the single schizophrenic women in this sample had had a serious relationship; 1 in 5 of them had a baby before hospitalization and 1 in 8 had a baby *after* admission. It is worth noting that a small group, 5 of the 813, each had had 4 or 5 illegitimate children. Stevens confirmed the a priori expectation that additional diagnoses of sociopathic personalities characterized those schizophrenic women with the highest illegitimacy rate.

The fertility of the mentally ill has attracted the attention of "do-gooders" and "do-badders" for more than a century; Philippe Pinel compiled data on the marital status of patients in French institutions in 1809, but the last half of the nineteenth century saw a growing, misguided fear that a dysgenic tide would flood society with both insane and

TABLE 15

Marital statuses of male and female schizophrenics in Zurich as contrasted with census normal controls

	Males (%)		Females (%)	
	Probands	Census	Probands	Census
Single	52	21	48	20
Married	27	73	18	63
Divorced	16	3	19	5
Widowed	4	3	15	12

Source: Adapated from Bleuler (1978).

retarded persons who had been abandoned by their families and who would multiply "like rabbits." Repressive legislation and the atrocities of the Third Reich were fed by such unfounded fears (Kevles, 1985; Lifton, 1986).

It was not until recent decades that unimpeachable facts on the fertility of psychotic patients became available. Once again we find it necessary to turn to the Norwegian social psychiatrist Ørnulv Ødegaard for an analysis of the reproductivity of all first admissions to mental hospitals from 1936–1955, a period before antipsychotic medications could affect release and remission rates of schizophrenics and those with affective psychoses. Table 16 shows that marital fertility is somewhat lower for schizophrenics compared to the general population and that both the marriage rate and the relative rate of reproduction (a product of fertility rate times marriage rate) are much lower, especially for male schizophrenics. The values for relative reproduction of 36 percent of normal for males and 48 percent of normal for females in this Norwegian population certainly suggest a lack of "Darwinian fitness" for carriers of the genotype for schizophrenia and, although not to the same degree, for those with affective psychoses.

L. Erlenmeyer-Kimling, a behavioral geneticist with long experience and expertise in the fertility of the mentally ill and its implications for human evolution, has confirmed the Norwegian picture of diminished Darwinian fitness in New York State. She has shown that the major mechanism of lower fertility rates is via lower rates of marriage. In a cohort of female schizophrenics admitted in 1936, only 49 percent had ever been married and, by 1961, their rate had increased only to 54

TABLE 16

Darwinian fitness in Norwegian psychiatric patients as a percentage of the general population rates

	Schizophrenics		Affective Psychoses	
	Male	Female	Male	Female
Marriage rate	38	53	91	93
Marital fertility	93	92	85	82
Relative reproduction[a]	36	48	77	76

Derived from Ødegaard (1972) for period 1936–1955.
[a]Fertility x marriage rate.

percent; for males, the values were 22 percent and 27 percent. For cohorts admitted in 1956, the ever-married rates in 1961 had increased to 64 percent for women and 43 percent for men, still very far from the rates in general population.

Gottesman and Erlenmeyer-Kimling (1971) organized a conference to discuss differential reproduction in individuals with mental and physical disorders. At that meeting, Eliot Slater, Edward Hare, and John Price, three British psychiatrists with long-standing interests in genetics and society, could report on the civil state and fertility of all inpatients and outpatients with various diagnoses who had been seen at the renowned Bethlem and Maudsley Hospitals during the years 1952–1966. Some of the important findings from the approximately 20,000 London-area adult patients are shown in Table 17. Given the conservative and valid diagnostic approach used in these hospitals (see Chapter 2), the data from some 1,000 schizophrenic women and 1,000 schizophrenic men are invaluable. The fertility rate of 2.2 children among *married* schizophrenics of both sexes was 95 percent of the rate among the general population of England and Wales at that time. The severe selection against reproduction of schizophrenics is revealed in Table 17 by information on children per patient regardless of marital status—0.9 per schizophrenic female and 0.5 per schizophrenic male; by the proportion ever-married—54 percent of women and 33 percent of men; and by the proportion of schizophrenic patients who had any children at all—only 39 percent of women and 22 percent of men.

Eugenic alarms and punitive legislation need not be invoked, according to the data reviewed here on the fertility and marriage rates of schizophrenics. However, there is a continuing need for empathy and compassion for the individual patient who becomes pregnant or has a baby, either intentionally or through failures in contraception or the predatory behavior of unscrupulous males. Continuous tracking and updating of secular trends in the marital and fertility patterns of psychiatric patients will inform both the social and the biological components of social biology and, in the process, will promote humane and rational social policy decisions about the care of the mentally ill.

Genetic Counseling

This is an appropriate place to broach the topic of genetic counseling — the use of genetic information to influence decisions about marriage, divorce, childbearing, and abortion — after reminding ourselves that the road to Hell is paved with good intentions and that we are dealing more

TABLE 17

Marital and fertility data, London patients, 1952–1966, to highlight the reproductive disadvantages for schizophrenics

Marriage and fertility groupings	Females			Males		
	Schiz.	Manic-dep.	Neurosis[a]	Schiz.	Manic-dep.	Neurosis[a]
Number of patients	1,086	2,692	5,596	1,003	1,606	3,902
Number of children[b]	907	3,715	6,397	452	2,218	4,168
Children per patient	0.9	1.4	1.1	0.5	1.4	1.1
Children per marriage	1.7	1.9	1.6	1.5	1.9	1.6
Children per fertile marriage	2.2	2.4	2.1	2.2	2.4	2.2
Proportion of patients ever married (%)	54.0	79.1	73.6	32.7	79.1	69.6
Proportion of childless marriages (%)	24.9	20.1	23.4	27.7	21.5	25.1
Proportion of patients with children (%)	38.9	61.1	54.0	21.8	58.2	49.4

Derived from Slater, Hare, and Price (1971).
[a]Excludes obsessionals.
[b]Includes illegitimate children.

with an art than a science for the time being. The decisions here are very personal ones, going to the core of an individual's sense of identity, and they are often made at the height of vulnerability to both competent and incompetent advice.

Inspection of the relevant journals and annual professional meeting workshops reveals that very little genetic counseling for psychiatric syndromes actually takes place. Clinicians who deal with psychiatric patients are seldom informed about the literature on psychiatric genetics, and counselors who specialize in medical genetics seldom are informed about psychopathology — there seems to be a "counseling gap" for psychiatric disorders that do not show neat Mendelian segregation ratios of 50 percent (Huntington's disease) or 25 percent (siblings of those with cystic fibrosis or PKU-mental retardation). As mental disorders become the focus of the media, a growth industry in genetic counseling for psychopathology may be in the offing. Such self-help consumer groups as the National Alliance for the Mentally Ill (NAMI) have expressed a keen interest in the issues surrounding the possible genetic transmission of schizophrenia, affective psychoses, and Alzheimer's disease, and NAMI has commissioned an introductory pamphlet on genetic counseling for the families (Gottesman, 1984).

Much of the information available to patients and their relatives is really misinformation. The well-known prototype of Huntington's disease, where remarkable advances have taken place in locating a gene at the tip of chromosome 4 and that can be detected with enough accuracy to identify carriers before they show symptoms, provides an inappropriate, even misleading, model for counseling in schizophrenia. Unnecessary guilt and self-limitation in both marriage and reproduction are the result of such misinformation. From the research reviewed in the previous chapters, it should be clear that environment plays an important role in risk potentiation and that the risks are quite variable in a group such as "the children of schizophrenics." Adoption authorities can take both the bad news and the less bad news captured in Table 12 (page 142) on the effects of keeping versus adopting children born to severely ill mothers. Both professionals who refer schizophrenics or their relatives for counseling and lay persons tend to carry much stronger beliefs in the magnitude of the "genetic effect" than are justifiable.

Ideally, genetic counseling should be dispensed upon request only and not legislated or imposed in a coercive fashion. But given the realities of ignorance and misinformation, professionals need to take more initiative in making the existing empirical information about the risks of developing schizophrenia available for rational genetic counseling. A leading question in regard to marriage intentions or to having kids to either or both relatives and patients is a good icebreaker. However, if a lead, once repeated, is not pursued by the client, it is best to drop it temporarily to avoid damaging any evolving "therapeutic alliance"; hopefully, a seed will have been planted for future harvesting. The major goals of genetic counseling should be to relieve the suffering of individual patients and their families, while at the same time protecting their rights as citizens. Such goals may need to be implemented by ombudspersons tutored in the state of the art of the genetic aspects of *common* genetic disorders. Early identification of vulnerable genotypes, and even prenatal detection, are in the realm of current possibilities, thereby increasing the range of ethical and social questions faced by the counselor. The temptation to play God or to impose one's own fears and values about schizophrenia are ever present and must be resisted, perhaps via peer consultation for second opinions.

Both the art and the complexities of genetic counseling for a complex characteristic such as schizophrenia can begin with a review of the risk figures shown in Figure 10 (page 96) for the different kinds of relatives and then concentrate on the values shown for the children and siblings of schizophrenics. The average risk for children is 13 percent, but when both parents are affected, the risk jumps to 46 percent. For siblings, the risk averages 9 percent, but jumps to 17 percent if, in addition to a schizophrenic sibling, one parent is also affected. A good deal of variability is masked by such average values, and they must be individualized before they can be of use to particular clients seeking information about their own family's situation.

An important clue to such individualizing was provided by Franz Kallmann in his 1938 monograph on the risks to Berlin families during the first half of this century. Briefly, the risks to the offspring of severely ill hebephrenic and catatonic schizophrenics were about 21 percent, but they dropped to 11 percent if the parents had mild or late-onset schizophrenias (called simple or paranoid); for all schizophrenia varieties combined, the risk to the children was 16 percent. However, when the risks were analyzed to allow for the mental and civil status of the coparent, more variability was uncovered. For example, children of mothers across all severities of schizophrenia who were born out of wedlock (implying, perhaps, abnormal personalities in fathers) had a risk for schizophrenia of 27 percent. At the other extreme, if the parents' schizophrenia was mild and the parents were married to normal mates, the risk to the offspring of such couples was only 2 percent. How then to further individualize the risk odds to accommodate the number of siblings, parents, and other relatives, those both affected and unaffected with schizophrenia, in any particular pedigree?

The computer program RISKMF, developed by the animal-breeding geneticist Charles Smith of Edinburgh, Scotland, for use with multifactorial traits, could be brought to bear on this difficult task. Each family could then have its own particularized, tailor-made risk assessment. A sample of the results that might be expected is given in Table 18. The results obtained would then become a point of departure for the remainder of the counseling process, still very much an art despite the use of the computer program.

With a multifactorial trait such as schizophrenia, the more healthy relatives one has and the closer they are to one genetically, the lower the risk. For example, with neither parent affected and two sibs already schizophrenic, the risk to the next sib (or to the next child of the unaffected parents) is about 14 percent. By adding one healthy sib to such a pedigree (two sick, one well), the risk to the next sib is reduced to 13 percent; by adding another healthy sib, the risk drops to 12 percent, and so forth.

The risks for multifactorial traits are modified in an important way when relatives on both sides, paternal as well as maternal, are affected with schizophrenia. Examples are given in the bottom half of the table. For example, in the second pedigree shown in the left column, with one sibling, plus one second-degree relative (such as an aunt or uncle or grandparent), affected *and* and with one parent also schizophrenic, the risk predicted for the next sib is either 22 percent *or* 28 percent, depending on whether the aunt who is schizophrenic is on the same side of the pedigree as the schizophrenic parent (22 percent) or as the normal parent (28 percent). The paradoxical increase in risk can be traced to the fact that the genotype of the normal parent is not really "clean" with respect to the genes for schizophrenia; this hidden fact is revealed when schizophrenia is observed in that parent's sibling, the aunt to the client of interest. In other words, the client is receiving relevant genes from *both* sides of the family tree; each side of the family pedigree contributes independent, additive risks to the client. Note that, in certain pedigrees, the predictions of risk are quite different from what they would be if one major dominant gene were all that was involved. Given two sibs affected but both parents normal, the risk to the next sib is 14 percent; given both parents affected but two sibs healthy, we can predict a 33 percent risk to the next sib. Two affected first-degree relatives in each pedigree generate quite different risks, depending on which relatives they are. If a major dominant gene were involved, as in Huntington's disease, a constant risk of 50 percent in all siblings and a constant risk of 75 percent in all offspring of two HD patients would be predicted. The vast range of risks tabled for just these few instances of pedigrees encountered in the real

TABLE 18

Recurrence risks for schizophrenia with varying family pedigrees theoretically derived

Pedigree	Number of Schizophrenic Parents		
	None	One	Both
No sibs	0.9	8.5	41.1
1 sib *unaffected* (U)	0.9	7.6	36.5
1 sib *affected* (A)	6.7	18.7	45.9
1 sib A + 1 sib U	6.2	16.6	41.9
1 sib A + 2 sibs U	5.5	14.8	38.9
2 sibs A	14.5	27.8	50.6
2 sibs A + 1 sib U	13.3	25.0	46.4
2 sibs A + 2 sibs U	12.0	22.4	43.4

	Same side — Opposite side of pedigree as affected parent		
1 2nd-degree relative A	2.7	10.6	19.0	45.3
1 sib A + 1 2nd-degree A	10.3	21.5	28.3	50.5
1 sib A + 1 2nd-degree A + 1 sib U	9.4	19.0	25.4	46.0
2 sibs A + 1 2nd-degree A	18.6	30.8	35.7	54.9
2 sibs A + 2 2nd-degree A + 1 sib U	19.9	30.0	38.0	54.1
1 3rd-degree relative A	1.7	9.6	13.3	42.8
1 sib A + 1 3rd-degree A	8.6	20.2	23.8	48.3
1 sib A + 1 3rd-degree A + 1 sib U	7.9	17.8	21.2	44.0
1 sib A + 1 2nd-degree A + 1 3rd-degree A	11.9	23.1	32.3	52.9
2 sibs A + 1 2nd-degree A + 1 3rd-degree A + 1 sib U	18.5	28.9	35.5	52.3

Note: Risks estimated assuming a lifetime risk of 1% in the general population and a lifetime risk of 10% in sibs of schizophrenics (correlation in liability of .40) in RISKMF computer program (Smith, 1971). Adapted from Gottesman, Shields, and Hanson (1982).

world highlights the complexities of genetic counseling. Self-counseling is not recommended.

Before leaving the topic, it is important to mention other helpful advice for the families of schizophrenics. One path to economic survival is to get as much noncancellable health insurance with the maximum psychiatric-care coverage that is feasible in a high-risk situation. Another is to take a firm stand in warning the families about their apparent vulnerability to certain drugs as triggers of schizophrenia. Many of these drugs have specific dopamine (a biochemical transmitter of nervous system signals) stimulating properties and include marijuana, cocaine/crack, PCP, amphetamine/crank, and LSD. Better safe than sorry — some vulnerabilities to schizophrenia might never be potentiated in the absence of exposure to such substances; individuals at risk must be told in no uncertain terms that they do not get a second chance to undo the precipitation of genuine schizophrenia by one exposure to crack, crank, or PCP (Tsuang, Simpson, and Kronfol, 1982; Andreasson et al., 1987).

Certain experiences should also be avoided by those predicted to be at moderate to high risk for schizophrenia. Extreme sleep deprivation is known to trigger psychoticlike behavior and may be a releaser of schizophrenia. The proliferation of cults, encounter groups, sensitivity training to "get in touch with yourself," and similar "mind-expanding" invitations must be avoided by those at a substantial risk of developing schizophrenia. In regard to family planning and childbearing, a conservative approach is recommended for schizophrenics and for those at high risk, not because of a fear of dysgenic consequences to unborn children, but because the added environmental stress of being a parent may have negative consequences for the course of the illness or the predisposition to it.

Legislative Constraints: Marriage, Immigration, and Sterilization

Few people would question the legitimacy of the government intruding upon our private lives to legislate the fluoridation of our drinking water or to require vaccination against such infectious diseases as smallpox or polio, or even to forbid us from spitting on the sidewalk. We do not question the prohibition of marriage between fathers and daughters or between brothers and sisters; however, 20 states do *not* prohibit first-cousin marriages, although such matings will markedly increase the rate of hidden recessive diseases. What should be the role of government in legislating about such private decisions as marriage and reproduction among those with major mental disorders?

By the beginning of World War II, 41 states had enacted laws to prevent the marriages of schizophrenics, other psychotics, and the mentally retarded. The assumptions behind these laws included the *belief* that such disorders were uniformly transmitted genetically, that contracts made by such persons were invalid, or that such persons would make incompetent parents and their offspring would become financial burdens to the state. The historian, Mark Haller, has noted that such laws appear not to have been enforced.

In half the states, as of 1989, "insanity" does provide grounds for divorce or annulment of marriage. However, some states require that the marriage must have lasted for up to seven years before permitting divorce on these grounds (Connecticut, after five years; Texas and Pennsylvania, three years; New York, Ohio, Massachusetts, and Illinois disallow psychoses as grounds for divorce). The authority to regulate such private and seemingly protected behaviors as marriage and divorce derives from the police powers in the Tenth Amendment to the Constitution, the same authority that allows quarantines and compulsory vaccinations for infectious diseases.

A legislative strategy devised by the United States Congress to control the economic burdens associated with the care of the mentally ill and retarded has been to refuse entrance to new immigrants who fall under these categories. Since many of the early advocates of eugenics also were enthusiastic about restraints on immigration, especially of nationalities different from their own, it is likely that the motives were not merely economic ones. In 1882, a law was passed that explicitly excluded lunatics, idiots, and persons likely to become public burdens. In 1952, the currently effective immigration law was enacted, which includes a much larger list of persons not welcome on our shores. Because they fall under the category of the "insane," schizophrenics are excluded from entering the country; also excluded are the mentally retarded, anarchists, drug addicts, psychopathic personalities, homosexuals, and other groups — a total of 31 categories. Furthermore, immigrants who become mentally ill or "insane" within five years of entry can be returned to their homeland under this law. The law is not enforced in a systematic way, but it is enforced; in reviewing many hospital charts in both the United Kingdom and Scandinavia, one will find the notation that a patient has been returned from the United States as a consequence of the mental illness exclusion provisions of the immigration law.

Compulsory, eugenic sterilization of the mentally ill and retarded is a practice most readers would associate with some kind of totalitarian government. At one time or another, however, 30 states had passed legislation authorizing compulsory sterilization. Most of the 22 states with such laws still on the books restrict their application, if ever, to institutionalized patients with mental illness or retardation, epilepsy, hereditary criminals (sic), sex offenders, and syphilitics. Indiana was the first state, in 1907, to legalize compulsory sterilization of inmates at the state reformatory by Dr. Harry Sharp, who had developed the technique of vasectomy as a replacement for castration (Reilly, 1985; Robitscher, 1973). (Such advanced and democratic countries as Denmark, Sweden, and Norway did not pass their humane sterilization laws until 1929.)

By 1964, 64,000 persons in the United States had been sterilized for eugenic reasons, half of whom were mentally ill. California alone was responsible for 20,000 of the sterilizations, traceable to the enthusiasm of one man who happened to be, first, head of the state lunacy commission and, later, superintendent of state hospitals. Currently, the practice of compulsory eugenic sterilization for mental illness or retardation has all but disappeared, and for good reasons: It not only offends our sensibilities about civil rights and liberties, but also is ineffective as well as unnecessary on scientific genetic grounds (see Tables 15-17). The number of involuntary sterilizations are dwarfed by the number of voluntary sterilizations for contraceptive purposes by married couples in the United States — 11 million couples (Statistical Abstract, 1988) are surgically sterile. It is the idea of *compulsory* sterilization that gives civil libertarians and the American Bar Foundation understandable pause; they have urged the repeal of all compulsory laws and a case-by-case review for those mentally ill or retarded or for their guardians who wish to avail themselves of such protection.

In the famous case of *Buck v. Bell*, in which the Supreme Court upheld the constitutionality of the sterilization of the mentally disabled, the superintendent of the Virginia State Colony for Epileptics and Feebleminded had invoked the health of patients and the welfare of society as his justification for wanting to sterilize Carrie Buck, a 17-year-old retarded female. Carrie's mother was retarded and promiscuous; Carrie was placed for adoption at age 4 and became pregnant (out of wedlock) at 17, giving birth to a baby who, *at age six months*, was alleged to also be mentally defective. Carrie's daughter died in 1932 from a physical illness after completing the second grade; she was described by her teachers as very bright (Kevles, 1985). The Virginia courts upheld the right of the state to sterilize Buck, and the decision was appealed to the United States Supreme Court. In 1927, on a vote of 8 to 1 approving sterilization, Justice Oliver Wendell Holmes, writing for the majority, was able to say, "Three generations of imbeciles are enough." Legal precedent in this case was traced back to the 1905 Massachusetts law requiring compulsory

vaccination against smallpox; it was reasoned that compulsory sterilization was analogous to compulsory vaccination against pregnancy for
those who would not refrain from getting pregnant.

The Nazi "Solution"

Unknown to anyone in the United States during the heyday of eugenic
enthusiasm, the stage was being set for one of the major atrocities of all
time — the systematic medical killing of mentally and physically disabled
patients and of "non-Aryan" peoples under the jurisdiction of Hitler's
Third Reich — all in the service of satisfying the paranoid, megalomania-
cal drive to purify the "Aryan" gene pool. Instead, it resulted in a moral
cesspool. Each of the six steps culminating in the Final Solution has been
chronicled with painstaking scholarship by Robert Jay Lifton, Distinguished Professor of Psychiatry and Psychology at the City University of
New York in *The Nazi Doctors — Medical Killing and the Psychology of
Genocide*. Strong resolve is required to get through the mind-numbing
depiction of what Hannah Arendt called the "banality of evil."

The new Nazi government, in a hurry to heal the physical and economic health of Germany in 1933 after Hitler's inauguration, passed laws
compelling eugenic sterilization. Judicial safeguards, cynically cosmetic,
required a review by a Hereditary Health Court of everyone, institutionalized or not, known by a physician to have a "hereditary" condition. The
following compulsory sterilizations were approved: mental retardation,
estimated at 200,000; schizophrenia, 80,000; manic-depression, 20,000;
epilepsy, 60,000; Huntington's disease, 600; hereditary blindness or deafness, 20,000; congenital malformations such as clubfoot and cleft palate,
20,000; and hereditary alcoholism, 10,000. The estimates sum to a projected total 410,000 sterilizations mandated to prevent what the Nazis
termed "life unworthy of life." At least 350,000 sterilizations were carried
out under the legislation, 56,214 in the first 12 months — twice as many
as the entire United States had performed during the preceding 30 years!
The court reviewing each case consisted of a lawyer and two physicians,
one of whom was knowledgeable about medical genetics. Only 13 percent
of petitions for compulsory sterilization were denied in the first year.
Between 1934 and 1945, 3.5 million men and women were sterilized on
"eugenic" grounds.

In October 1935, the laws for sterilizing the mentally and physically
afflicted could be characterized as harsh and as guided by fanatic political
ideology, rather than by any science of human genetics, but they were not
directly genocidal or anti-Semitic, and world opinion was not aroused in

protest. At this point, the sterilization laws were renewed and extended to marriage control via the "Law for the Protection of the Genetic Well-being of the German People." Violators were subject to imprisonment and their marriages nullified. The marriage application form became a searching cross-examination and the basis for a national register of "genetic" conditions; it demanded information extending to great-grandparents and first cousins and included their mental and infectious diseases, religion, race, criminality, and special "gifts" (e.g., noteworthy abilities in athletics, art, music, mathematics, etc.). Clearly, the earlier list of conditions leading to sterilization was being escalated for marriage control and now included hysteria, homosexuality, suicidal tendencies, mood changes, compulsive symptoms, juvenile diabetes, and hemophilia, to name a few (see Slater, 1936).

Even more shocking was the edict that phenotypically normal individuals whose relatives had any of the putatively dysgenic conditions could be forbidden from marriage and, henceforth, marriages between Jews and "Aryans" were illegal; anyone marrying abroad to avoid the laws could be imprisoned. A contemporary, eyewitness account of the first two steps toward the Holocaust is provided by Eliot Slater, who was a postdoctoral fellow in Rüdin's Munich Institute; in summary, he observed, "The Fuhrer directs with a series of ukases. With successive hammer-blows the German citizen is driven into a swastika-shaped hole. The atmosphere of compulsion pervades the whole of his life. The fact that he and fellow men are now to be selected and bred like a herd of cattle seems to him hardly more distasteful than a hundred other interferences in his daily life. . . . The command now is to breed" (1936, p. 292).

The third step in the program of racial purification would elicit the phrase, understated in retrospect, "calculated cruelty," from Justice Robert Jackson at the Nuremberg War Criminal Trials. Mercy killing or euthanasia on a case-by-case basis with strong ethical, legal, and medical oversight has a recognized role in the relief of intractable suffering and in the removal of life-support systems for the "brain dead," but it is understandably controversial. Such activities must be kept separate in our thinking from what is about to be described. "Mercy killing" of the mentally ill was discussed in the inner circle of Hitler's advisers as early as 1935 (Lifton, 1986). In a 1939 test case of a severely retarded and malformed child, Hitler authorized the euthanasia after the child's father had allegedly petitioned the physician in the case. At first, only infants and institutionalized children under age three were designated for medical killing after review, *of course*, and unanimous approval by a panel of three physicians. An overdosing with sedatives was the primary method of killing. All death certificates were systematically falsified to give the impression that death had been caused by something such as pneumonia to prevent the truth from being revealed to the parents or to the public. The age limit was soon abandoned, and it is estimated that at least 5,000 institutionalized retarded children were killed in the program.

The fourth step is of most relevance to the schizophrenia story and took place in the fall of 1939 with Hitler's decree (written on his private stationery) that "patients considered incurable" could be granted a mercy death (*Gnadentod*). No public law was actually passed for killing either child or adult mental patients, to deny a propaganda advantage to the outside word; the actual killing was kept secret. The code name "T4" was used for the program to kill the much larger numbers of persons with schizophrenia, epilepsy, senile dementia, organic psychoses, and Huntington's disease; those continuously hospitalized for five years; the criminally insane; and other patients *who were not of German blood,* — that is, Jews and Gypsies. The magnitude of the program must have required the involvement of most psychiatrists and many other professionals. The proper forms (Lifton, 1986) had to be filled out and, again, a committee of three experts as well as a "senior expert" had to approve the medical killing.

Six main killing centers were set up at converted mental hospitals. The volume of killings required research to determine the most efficient method; after a number of trials, Hitler, on the advice of Dr. Werner Heyde (a leading psychiatrist), selected carbon monoxide as "the most humane" method of killing. It is estimated that 100,000 psychiatric patients were murdered in this program.

As in the case of children, death certificates for the schizophrenics and other patients were falsified by physicians to deceive the families and the public. The dead were cremated and ashes were sent to the families. As the shooting war had begun in September 1939, the wartime mentality engendered might have been used by the personnel involved in the killings to assuage their consciences. In line with the broader theme of this chapter, the rest of the story about Nazi *dysocial* policy must be told.

From April 1940, Jews in institutions, as well as Jews and other "non-Aryans" in concentration camps within Germany, were moved to the killing centers (code name 14f13) and exterminated; an estimated 21,000 persons all told by the end of 1941 — the fifth step. The genocidal mentality was now safely in place, as was the necessary killing machinery. A new, quicker poison gas had been developed, hydrogen cyanide, for the sixth step, the Final Solution to purifying the Aryan race. The Jews remaining in Germany were exported to occupied Poland where, together with Jews from the conquered territories such as Poland, Hungary, the Netherlands, France, Czechoslovakia, and Greece, they were annihilated, either immediately or after being exploited as slave laborers. Altogether, from late 1941 to war's end, an estimated 4 million

Jews were murdered in the camps, a further 2 million in the course of the war, and a further estimated total of 4 million non-Jewish, non-Aryan civilians.

The Nazi killing frenzy to destroy allegedly bad genes, whether they existed in schizophrenics, Jews, Gypsies, communists, diabetics, or retarded children, aided and abetted by scientists and physicians, is clearly one of those unspeakable atrocities that must be given a voice and a hearing. Those of us engaged in the search for the causes of mental disorders such as schizophrenia cannot safely assume that our efforts will be unambiguously distinguished from the "insane" practices of the past. Inconvenient as it may be at times, incorruptible watchdog committees of informed ethicists, scientists, physicians, patients-in-remission and their relatives, jurists, and ombudspersons are needed to bolster the collective conscience of society. The street connecting society, social policy, and scientific schizophrenia research must be kept free of careless and reckless leaders, citizens, and scientists.

The jurist B. M. Dickens of the University of Toronto Faculty of Law, after reviewing global legislation on eugenic sterilization and selective abortion in the light of their histories and the promise of gene-replacement therapy, reached the following sage conclusion worthy of endorsement by any ombudspersons for society *and* science:

The challenge is to identify the costs both of freedom and of control, and to decide by political processes how far to bear the costs of individual freedom on society, and the costs of social control on individual liberty and responsibility. (1987, p. 682)

The case history that ends this chapter was selected to illustrate the frustration that parents of schizophrenics experience in regard to the social policies on imposing treatment and involuntary commitment.

A Parent's View on Enforcing Medication
One of the founding mothers of a support group expresses her frustration with the law's positions on involuntary commitment and preventive treatment for her daughter.[1]

---

[1]E. Slater, *Schizophrenia Bulletin*, 1986, 12: 291–292.

My daughter developed paranoid schizophrenia over the last 15 years. Two years ago, she stopped taking her medication and began sliding back into a familiar pattern of delusions and accusations. She had begun making frantic calls to the police and was about to be evicted from her apartment for harassing the other tenants when she physically attacked her fiance and was taken to a community hospital in her catchment area. At the recommitment hearing after the initial 5-day observation period, she was disturbed enough to be held involuntarily for another 20 days.

After the hearing, her fiance and I spoke to a psychiatrist and were shocked to discover that the community hospital did not force patients to take medication. The psychiatrist said they would talk my daughter into taking it. From 13 years of experience with her problem, I knew they couldn't convince her, in her disordered state, to take medication. Luckily, we were able to find another community hospital that would make sure she got the medication she needed, and we reached an agreement to have her transferred there.

Watching someone you care about disintegrate over a period of months into someone you can't recognize, someone made fearful and frustrated by her own imaginings, is painful. To be forced, as her fiance was, to have a loved one committed because of physical violence is deeply traumatic for everyone involved. The most disheartening thing was finding out she wouldn't get the help she needed at the end of all that.

I was so upset by our experience that I wrote to the Commissioner of Mental Health for Pennsylvania, questioning the policies of this particular hospital. He replied that he thought a patient committed by the court for treatment should be treated and that if there was a question about the advisability of forcing medications, a second doctor should be consulted. His office, however, could not interfere with a community hospital's policies.

I was surprised to learn that many States do not force medication. I would guess this policy might be the result of hospitals being sued by patients or patients' rights groups. The families have been so frustrated by this very serious problem that they have sometimes even considered taking legal steps to see that their family member was effectively treated with medication.

As a regular policy, not forcing psychotic patients to take medication would seem to create many problems for hospitals. How do hospitals deal with patients who may harm themselves or others? Should they be confined in restraints, as they were years ago? Should they be kept in isolation, which requires large facilities? Or will they end up being discharged back into the community as untreatable?

One of the things that our support group, Parents of the Adult Mentally Ill (PAMI), does is try to implement sensible treatment based on our own years of experience with mentally ill children. Most, if not all, of us tried for a long time to talk our sons and daughters out of their delusions before we came to accept that reasoning just doesn't help. In the same way, we know how impossible it is to convince someone who is mentally ill to take medication when the person is becoming irrational. The problem isn't one of intelligence; most of our children have some college education and some, like my daughter, have college degrees. The problem is in the nature of the illness itself. According to Dr. E. Fuller Torrey (1983)[2] of St. Elizabeths Hospital in Washington, D.C.:

People with schizophrenia have diseases of the brain; they cannot give informed consent. . . . Supporting this position is a study of chronic schizophrenic patients showing that only 27% of them understand that they needed medication. [p. 191]

As a parent, I also know that medication is not perfect and that the side effects can be distressing. When my daughter goes back on her medication, I feel bad seeing her shuffling or experiencing involuntary arm and mouth movements. These symptoms usually subside over time; but she also gains weight, and she hates being heavy. I think she hates taking medication most of all because she is, in a sense, admitting she is mentally ill, something she very much wants to deny.

But the alternative is much worse. She becomes more and more paranoid and delusional. I can only imagine what hell her life must be when she tells me people are trying to kill her, or that she knows I'm trying to commit suicide, or that evil people won't let her sleep at night. I can see her distress as her world begins crashing in or her and she needs help. I don't think she needs the right to be crazy or that she, or any other mentally ill patient, needs the right to commit suicide or act in a way she will later regret.

One thing that some family support groups have done to try to balance the need for medication with its adverse side effects has been to launch a letter-writing campaign to drug companies urging them to spend more time and money on research. I understand that some of the newer medications have been improved. On the other hand, while many parents think hospitals should enforce giving medication to mentally ill patients who have had to be committed to their care, we would also like to see hospitals question or eliminate the practice of giving massive doses of medication immediately on admission. This can cause patients to fear

---

[2]E. F. Torrey, *Surviving Schizophrenia: A Family Manual*, New York: Harper & Row, 1983, pp. 182–195.

the medications. Although this is done less frequently, drugs should be used as discriminatingly as possible to help the patients.

My daughter has been hospitalized a number of times over the last 15 years, and it takes at least a month for the medication to begin to turn her thinking around. Then she begins to show a remarkable change in her thinking and in her personality. She becomes a sweet and loving person, reminiscent of the daughter I used to know. If a proposed ruling countermands enforcing medication in this State, if frightens me to think that she and thousands of others may have no help when they become irrational. I don't want to see that wasteland of frustration and torment become a way of life for her, and for thousands like her.

11

The Big Picture
Interactive Synthesis and Integration

The parable of the blind men describing an elephant from each of their own limited vantage points and the consequent humorous lack of congruence with the reality that sighted witnesses would report has often been applied to schizophrenia research —with good reason! Even sighted observers may be confused if they lack experience and a frame of reference for integrating their impressions. One of my favorite jokes is about the naive provincial farmer who telephones the sheriff after an elephant has escaped from a traveling circus to report that a huge, headless monster is pulling up cabbages from the garden with its tail. In reply to this question, "What is it doing with them?" the farmer says, "You wouldn't believe me if I told you."

No one or two scientists laboring in their laboratories or over their computers in one or two disciplines, even after consultation with gifted social workers, psychologists, nurses, psychiatrists, and relatives of patients, would have a sufficient sample of the truly relevant elements to get the big picture. Consider, as an analogous task, the elements required to land a team of astronauts on the moon, have them conduct their ultimate scientific mission successfully, and return them to our planet with information that will contribute to a deeper understanding of the origins of the universe, once that information is integrated with a preexisting matrix of solid, soft, erroneous, and missing facts. Impossible as that mission would have sounded 30 years ago, it has already been accomplished. What has not been accomplished, obviously, is the parallel mission to complete the big picture for understanding the genesis of schizophrenia.

But there are reasons for optimism. Only now, as we near the end of the twentieth century, are the necessary elements finally available for a comprehensive understanding of schizophrenia. At the beginning of the century, Kraeplelin and Bleuler provided detailed phenomenological sketches of dementia praecox/schizophrenia. Their sketches were strong on the descriptive psychopathology of the syndrome, emphasizing symptoms, onset, course, and outcome, and were embellished with notions about treatment and speculations about etiology or cause. Just as many different maps are needed to understand a continent — topographical, political, historical, meteorological, population density, dominant land use, and so forth — so too must many vantage points and content areas be integrated to understand the "continent" of schizophrenia.

Although many different fields of knowledge have been brought to bear on our understanding of the etiologies of schizophrenia at this point in the book, they all have in common a whole-person and whole-population, or "macro," approach. Omitted so far, but known to be critically important for an eventual synthesis of knowledge, are such fields as neurochemistry, neuroradiology (brain imaging), neuroanatomy, neurophysiology, cognitive psychopathology (attention and information-processing deficits), and electrophysiology — representing the "micro" and "inside-the-skin" approaches. Schizophrenia has been usefully and heuristically described as a "neurointegrative defect"; our failure as scientists and clinicians to comprehend schizophrenia adequately may be described as a "conceptuointegrative defect." An adequate understanding of schizophrenia demands the interactive synthesis of disparate fields of knowledge. The problem calls for a team approach like the one NASA used to take human beings to the moon and bring them back.

In a creative and curmudgeonly metadialogue composed by Joseph Zubin between himself and Emil Kraepelin (1987), we find the following to guide our attempts to integrate and synthesize:

J.Z.: Why does it take so long to make progress in the field of schizophrenia?

E.K.: Well, it may be the case that we "knew" more in the early part of this century than we "know" now. In the USA I once heard someone say "It ain't ignorance that causes all the trouble. It's knowing things that ain't so!" Perhaps we had to unlearn false knowledge before we could advance to the new, cut down the underbrush before the new plants could thrive. (p. 361)

The preceding 10 chapters reflect an effort to describe the facts, certain and uncertain, that inform a view of (1) the nature of predisposing factors to schizophrenia and how they may be transmitted within families, and (2) how diathesis (predisposition) or vulnerability interacts *epigenetically* with stressors to produce episodes of dysfunction frequently followed by degrees of improvement. In the process, attention has been paid to cutting down the underbrush and to unlearning false knowledge. Fundamental impediments to progress come from the difficulty of specifying the object of study — schizophrenia — in a reliable and valid manner. The impediments and the partial solution via a polydiagnostic approach — so as to avoid dealing merely with a legislated convention from any one "theocracy" — have been emphasized repeatedly in earlier chapters. But it is good to remind ourselves that the problem will not go away; we do not yet "possess an equivalent to the pathologist's and the microbiologist's report telling us the 'right answer' at the conclusion of a clinicopathological [autopsy] case conference" (Meehl, 1986, p. 222). Paul Meehl further reminds us that "nothing but dogmatism on the one hand, or confusion on the other, is produced by pretending to give operational definitions in which the disease entity is literally identified with the list of signs and symptoms. Such an operational definition is a fake" (1986, p. 222). The nature of the diathesis, the nature of the stressors when they exist, and the nature of the interaction are yet to be determined by a NASA-like team approach. The skills and knowledge contained solely within an organization such as the National Institute of Mental Health are not sufficient for the mission.

Models as Maps: Where Are We and Where Do We Need to Go?

Unbiased readers of this text will have to conclude that, based on the cumulative credible evidence, a large, rather specific, and important genetic factor(s), in conjunction with putative, unspecified nongenetic factors in most cases, leads to the development, over varying lengths of time, of varying severities of schizophrenia(s). Resistance to such a balanced conclusion, when it appears, must be based on ideological reasons. However, even well-respected scientists working on the periphery of their expertise can produce politicized tracts about the ways that human behavior may be influenced by genetic factors (see, for example, *Not in Our Genes* by Richard C. Lewontin, Steven Rose, and Leon Kamin).

Given the existence of an important genetic predisposition to the liability for developing schizophrenia, it becomes important to build theories or models for how that predisposition becomes actualized into episodes of overt schizophrenic behaviors and to build other models for how that predisposition may be transmitted across generations. Such models can serve as maps for both the specialist researcher and the informed layperson to the continent of schizophrenia, telling us where we are, where we need to go, and where there are uncharted territories or inconsistent data requiring further exploration. Preliminary orientation to simple recessive and dominant gene models, as well as to more complex multifactorial-polygenic-threshold models, was provided in Chapter 5. If the models are to do their job of making sense out of the data, they must be dynamic; that is, they must take into account the changes in phenotype over time from normal to disordered to social recovery and, too often, back to disordered. Schizophrenia is almost always an episodic mental disorder. Recoveries from such episodes were known even before the era of active neuroleptic treatments; then, as now, about 20 percent of the persons diagnosed as suffering from schizophrenia could be expected to resume useful lives outside the hospital (Ciompi, 1988).

At the present stage of our knowledge, it is more accurate to call schizophrenia a heavily genetically influenced disorder rather than a genetically determined disorder, to avoid any hereditarian implication of inevitability. Some human malfunctions, whether medical or behavioral, are better than others for providing useful clues to the kinds of research strategies into causes and treatments for schizophrenia that will be effective. Coronary heart disease, diabetes, mental retardation, and epilepsy — all complex, multifactorially influenced syndromes with demonstrable genetic components — should be at the top of a list of recommendations. The hemolytic anemia associated with eating fava (broad) beans was described earlier as a prototype for true genotype-environment interaction; without the X-linked recessive genotype leading to a particular enzyme deficiency, persons eating the beans could do so with impunity. Both the bean and the gene were necessary for the development of this disease. Another textbook example of a recessive inborn error of metabolism, galactosemia, leads all babies homozygous (having received one copy from each of two parents) for the gene to develop the disease when exposed to milk in their diets, a universal agent for babies. This rare condition, affecting 1 in 30,000 at most, makes it easy to identify as a *genetic disease*, even though it is elicited by the ubiquity of milk. We cannot say that milk causes the disease—that would make it an *environmental disease*—because of the disparity between the frequency of the genotype and the frequency of the eliciting agent.

As a prototype of a disease that deserves to be called environmental, we can look at measles. Virtually everyone exposed to the virus gets the symptoms if they have not been inoculated against it. Given the universality of the genetic vulnerability to measles and the episodic, acute exposure to the virus, we can easily say that such conditions are "environmentally caused." Family and twin studies would readily demonstrate the familiality of measles, but the very high concordance rates in both identical and fraternal cotwins of probands, as well as a conjectured very high concordance rate between unrelated stepsiblings or within an old-fashioned orphanage, would point to something transmissible but *not* genetic.

Schizophrenia, like favism, is both a genetic and an environmental disorder, but the relatively low prevalence of the genetic predisposition compared with the relatively high prevalence of the various environmental causes/triggers/risk factors leads us to attribute the fundamental causes of schizophrenia to genetic factors. Neither a genetic disease such a galactosemia nor an environmental disease such as measles, schizophrenia fits somewhere in between, along with diabetes and coronary heart disease. For the latter two conditions, the unspecified genetic predisposition must share the spotlight with an array of environmental contributors (diet, exercise, smoking, lifestyle, pregnancy, etc.), some of which lead some people with the genetic vulnerability to develop the disorder. With such middle-of-the-continuum conditions, the facts suggest that the environmental factors may be interaction effects only— that is, only those unlucky persons with "sensitive genotypes" feel the environmental effects.

Models for the Development of Schizophrenia: The Pathway from Genotype to Psychopathology

Let us postpone a discussion of the models needed to account for the transmission of schizophrenia across generations within families until we have examined some schemes that take the genetic predisposition as a given and then proceed to account for the unfolding of that predisposition—its psychopathogenesis—into full-blown dysfunction.

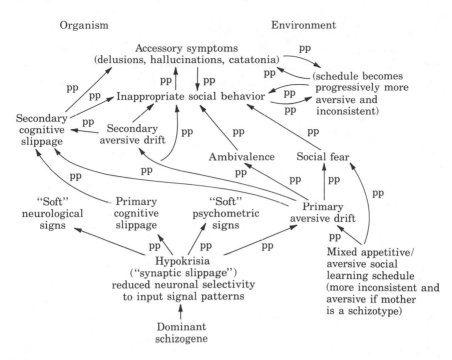

FIGURE 14. P. E. Meehl's speculative causal chain of events leading from a gene to the symptoms of schizophrenia by way of posited states of synaptic slippage, schizotaxia (inferred from neurological and psychometric signs), and schizotypy (inferred from a set of polygenic potentiators (pp); pp include anhedonia, social introversion, fragile body type, etc. (Courtesy of P. E. Meehl, 1966 and 1989.)

For many years, Paul Meehl has been advocating a partial causal network for the development of schizophrenia, which he calls one of "minimal complexity." It is reproduced in Figure 14 for its heuristic value, despite the challenging neologisms and jargon. Meehl sees the network as more heuristic than the basic causal chain of events he espouses, wherein a dominant schizogene leads to a hypothetical neurological state he terms *schizotaxia*, which leads to the development of a personality disposition — *schizotypy* — and then on to the development of the symptoms of schizophrenia proper in a subset of decompensated schizotypes. Figure 8 (page 90) already anticipated the concept of a liability toward developing schizophrenia that went beyond genetic liability at conception to allow for augmentation of liability by environmental sources and for reduction in liability by various assets. However, this model was a cross-sectional representation fixating on the end-state outcome and was not dynamic.

Meehl's flowchart map of the pathways or "causal chains" to the symptoms associated with diagnosing the syndrome of schizophrenia was formulated before the emphasis on biological explanations over the past two decades; precise definitions of the concepts used are still premature —he has earned the poetic license. The map encourages research on either organismic (the individual) or environmental variables—the two "reminders" juxtaposed on the map—at either macro or micro levels. He says:

Schizophrenia is a complicated collection of learned social responses, object-cathexes, self-concepts, ego-weaknesses, psychodynamisms, etc. These are dispositions of first or second order [more proximal causes]. They are *not* provided by our genes. They are acquired by social learning, especially learning involving interpersonal aversiveness. Assume the mutated gene (a structure) causes an aberrant neurohumor that directly alters [neuronal] signal selectivity at the synapse (Meehl, 1962). Then the gene is a *structure*; the gene-controlled synthesis of an abnormal substance (or the failure to make a certain substance) is an *event*; the altered synaptic condition is a *state*; and the results of that state's existing at the billions of CNS [central nervous system] synapses is an altered parameter of CNS function, i.e., a *disposition*. But this disposition is a disposition of at least third (perhaps fourth or fifth) order with respect to those molar dispositions that are the subject-matter of clinical psychiatry and psychoanalysis. Hence, an individual's being characterized by a certain genotype is a disposition of still higher order, because (presumably) the synaptic disposition itself is not an absolutely *necessary* consequence of the first link of the gene's action, since it could be avoided if we knew how to supplement the brain's inadequate supply of magic substance X, or how to provide a related molecule that would bring the parameters of CNS function back to the "normal" base. (1972, pp. 15–17.)

The model sketched by Meehl-as-artist is a multifactorial one— currently known in genetics as a "mixed model"—combining a mixture of a hypothesized dominant gene, plus many polygenically determined potentiators (shown as *pp* in the figure), plus interactions with the psychological environment, all leading to intermediate states and dispositions to the schizophrenic symptoms themselves. Although *anhedonia* (defect in experiencing pleasure) formerly appeared in the figure as a core element of schizophrenia, Meehl (personal communication, 1989) no longer considers it as other than one more potentiator (*pp*). Included on Meehl's list of contributing polygenic potentiators or protectors are the following: high primary social introversion, high trait anxiety, high or low trait aggression, low hedonic potential, low energy level, low mesomorphic toughness (a fragile body type), abnormal perceptual/cognitive abilities, such assets as brains and beauty, and so forth. By calling them polygenic, Meehl must mean that they each originate from a mixture of genetic and nongenetic elements. Reducing the terms in the map to actual neurophysiological or neurotransmitter pathways and their projections in the brain is a task not yet accomplished, but one begging for solution (Meehl, 1990). The top of the flowchart shows the symptoms of schizophrenia; the middle implies the schizotypal personality organization; and soft neurological signs, cognitive slippage, and psychometric signs at the bottom make up the key elements of schizotaxia. Meehl tells me that he believes that tests of attention and information processing may provide indicators of schizotaxia.

Another map of a causal network is oriented toward the conceptual clarification and explanation of the development of episodes of schizophrenia in individuals already known to be schizophrenic, but in a state of remission (see Chapter 7). Keith Nuechterlein, Michael Dawson, and Robert Liberman, specialists in information processing, psychophysiology, and rehabilitation in schizophrenics, at the University of California at Los Angeles, influenced by the causal chains/maps suggested by Meehl and by Gottesman and Shields (1972, 1982), devised the heuristic framework in Figure 15 as a guide to the possible prevention of relapse and the promotion of rehabilitation (Nuechterlein, 1987; cf. Goldstein, 1987). Their framework enlarges the perspective on developmental psychopathogenesis in a useful way and complements the earlier efforts.

Three stages are depicted in the figure: (1) the remission period for someone already known to be afflicted with schizophrenia, (2) the prodromal period when various vulnerability factors have already interacted with potentiators and stressors to produce intermediate states leading to prodromal symptoms (the symptoms associated with schizotypal personality organization and equal to the *pp*s in the middle of Meehl's flowchart), and (3) the stage of schizophrenia itself as the outcome of the process. By depicting other outcomes — social and occupational functioning — the figure implies subsequent cycles of the process without schizophrenia being the inevitable result.

As with Meehl's division into organismic and environmental variables and Gottesman's and Shields' division of contributors to liability into genetic and environmental assets and liabilities, Nuechterlein and colleagues separate out four classes of variables: personal vulnerability factors, personal protectors, environmental protectors, and environmental potentiators. No effort is made to divide each class into its antecedents, but laying out the causal chain in this fashion may have that heuristic effect. All such causal chains are preliminary; for example, the placing of "antipsychotic medication" under personal protectors in one scheme and under environmental protectors in another may be just a matter of "taste." Similarly, some personal vulnerability factors in Figure 15 may be subsumed under the intermediate states in Figure 14. "Busy" as it may

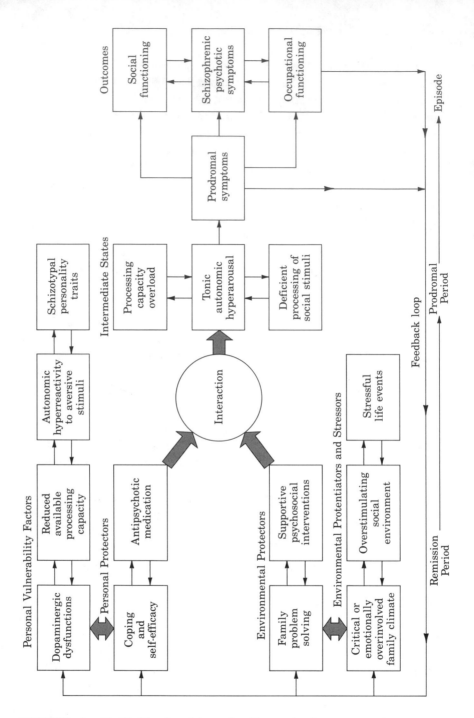

FIGURE 15. K. H. Nuechterlein and colleagues' schematic rendering of the complex interactions among hypothetical personal vulnerability factors, personal protectors, environmental protectors, and environmental potentiators and stressors required to account for changes in schizophrenic symptoms over the course of the illness. (Courtesy of K. H. Nuechterlein; cf. Nuechterlein, 1987; Goldstein, 1987.)

seem, Figure 15 still is far from as complex as it needs to be eventually; no biological paths are spelled out, and interaction must still be graphically finessed as a "circle" in the middle of everything else. The important cybernetic principle of *feedback* is recognized by the provision of a feedback loop, but no one can yet tell us what it is that is fed back.

Heinz Katschnig of Vienna makes the point that such "boxologies" (Figures 14 and 15) are heuristically valuable even though "boxes connected with arrows do not explain anything," because they sensitize researchers to the multifactorial nature of causal chains, thus promoting the integration and synthesis of divergent schools of thought. Figure 11 (p. 157) from EE research showing the joint effects of medication and life events on the threshold for generating schizophrenic symptoms is another good example to keep in mind.

Neither of the causal chain diagrams or models shown above facilitates the understanding of the *individual* cases of schizophrenia characterized in the various case history stories in Chapters 3 and 9 and elsewhere in this book. The reason is that each instance of schizophrenia arose from a unique constellation of the various elements depicted in the figures and described in the preceding chapters. This means that, eventually, each person with schizophrenia will require his or her own customized life chart in order for us to understand that person's dysfunction and to serve as a useful blueprint for intervention and rehabilitation.

In Figure 16, an attempt is made to show the unfolding of the schizophrenic processes in different, prototypical genotypes, making provision for the dynamic interactions between individuals and their experiences and between genes and their endogenous and/or external activators. Implicit in this and the models above is the notion of a *final common pathway* to some neurophysiological process, downstream from which the symptoms of schizophrenia eventually appear and upstream from which idiosyncratic combinations of the elements are sufficient to initiate the chain.

Implicit in the figure are elements from Meehl and Nuechterlein et al., as well as from Figure 8, with an additional threshold shown for the zone of schizophrenia spectrum disorders, and with the addition of a horizontal axis to allow for the dynamic changes in combined liabilities that arise across time. Such epigenetic interchanges between genotype and experience can lead to dynamic increments *or* decrements in the liability toward developing schizophrenia and can be revealed in this kind of "cartoon." The intention is to incorporate the concept of changes in "effective" genotype due to gene regulation (Vogel and Motulsky, 1986); that is, the switching on and off of gene activity by environmental inputs or by endogenous timing programs and by possible prenatal and postnatal critical periods indexing heightened (i.e., nonlinear) sensitivities.

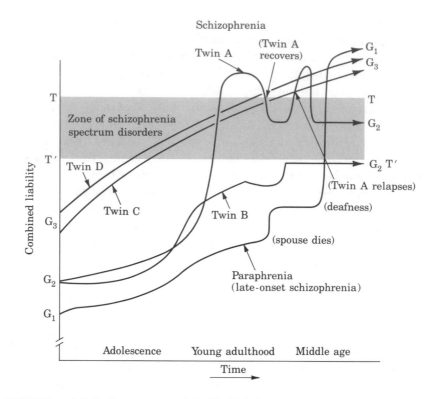

FIGURE 16. I. I. Gottesman and J. Shields' sketch of the hypothetical epigenesis (dynamic, reciprocal interactions between genotype and environment) of
schizophrenia over time, illustrated for a singleton (paraphrenic) and two pairs of
identical twins with varying initial liabilities. Adapted from Gottesman and
Shields (1982).

In Figure 16, the time axis clock begins when the sperm fertilizes the
ovum (in principle, it could start at the production of the two gametes
themselves), so that the impact of various prenatal and perinatal factors
—traumas, licit and illicit drugs, radiation, viruses, birth complications,
etc. — could be revealed on the vertical axis indexing combined liability
toward schizophrenia or spectrum disorders. Developmental changes associated with the processes of maturation, as well as random events (good
and bad luck), will influence the trajectories pictured for individuals
toward and away from the threshold. We would expect, given our emphasis on epigenesis, that the coming together of negative elements close in
time would have a synergistic and cascading effect on liability by not
allowing time for the natural processes of healing and repair to take

place. Thus, a predisposed young man — wrenched from his psychosocial support network, exposed to the rigors of basic military training, his fragile sexual identity threatened by "homosexual" horseplay in group showers, and fatigued from sleeplessness — decompensates when forced to make his way through strange terrain during nighttime maneuvers. He has quickly arrived at the intermediate states (processing capacity overload, etc.) in Figure 15 and the secondary cognitive slippage of Figure 14. Interruption of the chain of events or spreading them out in time — say, by a weekend pass — might have been enough to abort the passage to a psychotic episode. Conceptually, it is often difficult to draw a line separating "risk factors" from vulnerability factors; that is, those more intrinsic to the diathesis itself.

Three different genotypes and the trajectories of their changing liabilities over the life span are shown in Figure 16. No valid means is yet available for measuring levels of combined liability below the threshold for schizophrenia spectrum disorders; until that level is reached, we are dealing with "endophenotypic" information — data not available to the naked eye that are intermediate to the phenotype and the genotype for schizophrenia — various indicators with differing amounts of empirical support, such as measures of attention and information processing, subclinical personality traits, and the number of dopamine receptors detected in specific brain regions by PET scanning. G_1 traces the combined liability trajectory for a man who eventually receives a diagnosis of "paraphrenia" or late-onset schizophrenia in his middle 50s. He begins life with a low-average specific genetic liability to schizophrenia (the first axis in Figure 8) and average values for the other four liability and asset axes. Ordinary aging takes its toll, gradually increasing liability, and then he experiences the traumatic loss of psychosocial support by the death of his spouse in his 40s; a few years later, the onset of deafness with its psychological consequences adds enough liability to cross the T-T threshold and to precipitate an acute episode of schizophrenia of average severity.

A pair of identical twins born with an average specific genetic liability to schizophrenia is depicted by the two trajectories labeled G_2. Twin A experiences more perinatal stress than does Twin B and thus has more (endophenotypic) liability in infancy. Only Twin A experiments with hallucinogenic drugs in college and proceeds to an acute onset of schizophrenia in her early 20s; she receives adequate treatment and remits to a schizotypal personality or a "residual schizophrenia" (between the lower threshold T'-T' and the higher T-T); after ceasing her medication, she has another relapse followed by partial remission. Her cotwin accumulates liability, but never enough to exceed the lower threshold for a spectrum disorder, at least by age 50, but a hypothetical endophenotype-detector would show how close she is to becoming a diagnosable case. In the meantime, this pair would always be a discordant pair of identical twins in a low-technology research project and thus would diminish the magnitude of any demonstrated genetic effects.

Another pair of identical twins, C and D, is characterized by the trajectories labeled G_3, but this time the pair becomes concordant for schizophrenia in midadolescence with no detectable external sources for increments to their combined liability "account." Unluckily for them, they began life with a very high (D even greater than C) specific genetic diathesis for schizophrenia and drifted into a schizoid personality pattern (crossing the T'-T' threshold) and then insidiously into schizophrenia proper (crossing the T-T threshold).

A growing number of neuroscientists, encouraged by the failure to demonstrate credible stressors for the diathesis-stressor model of schizophrenia in many adolescent or young adult patients, have suggested that stressor absence may signify an endogenous fault in "programmed" synaptic elimination (Feinberg, 1982). (A synapse is the gap between neurons in a circuit; the circuit is completed by neurotransmitters such as dopamine.) Normal development in mammals involves a reorganization of the brain beginning at puberty that includes a reduction in the apparently surplus density of synapses. This loss in plasticity, it is speculated, is accompanied by an increase in specialization of the nervous system (e.g., sustained logical thought and complex problem solving). Certainly the amount of speculation far exceeds the amount of relevant data, but the promise of this area of research for closing the gaps in knowledge about the development of schizophrenia makes the ambiguity tolerable. Errors in the programming for normal "pruning" of neurons could involve either a delay in starting or a delay in ending. Saugstad (1989) suggests that schizophrenia results, in part, in late-maturing children (delayed puberty) "in whom greater than optimal neuronal pruning has resulted in excessively reduced synaptic density" (p. 37). She explores a wide range of epidemiological and demographic observations that are compatible with this speculation. A more general discussion and review of the neurobiology of development has been provided by Mary Carlson, Felton Earls, and Richard Todd, with a rich bibliography.

Such pruning-error hypotheses (Hoffman and Dobscha, 1989) can be seen as complementing the ideas embedded in the three causal chain models already presented and not as a replacement for them. The idea is certainly congruent with the epigenetic version of diathesis-stressor constructions. Hopefully, putting all three sketches or cartoons up on the wall simultaneously will stimulate new hypotheses and further progress by refutation of some and confirmation of others.

Gene-Environment Models for the Transmission of Schizophrenia

Models for the development of schizophrenia can exist independently of considerations about the models needed to account for the transmission of schizophrenia across generations and within families. As noted throughout this book, the difference is between pathogenesis and predisposition. Only a general treatment of models for the transmission of the predisposition to developing schizophrenia is given here. To go further would require technical and tedious discussions that might highlight the trees but obscure the nature of the forest (see Faraone and Tsuang, 1985; Lalouel et al. 1983; McGue, Gottesman, and Rao, 1985; McGue and Gottesman, 1989). Different models have different implications for the strategies that are pursued in etiological research, for the mechanisms of molecular dysfunction that may be involved, for rational treatment and prevention, for the detection of cases premorbidly, and for the utility of genetic counseling. It is known, for example, that diseases transmitted as recessive disorders are associated with one or another enzyme deficiency; however, of the 750 definitely identified recessive conditions known in our species by their patterns of inheritance, only 250 have had the aberration in an enzyme identified. If a missing enzyme should ever be discovered in some subgroup of schizophrenics, it would motivate a search for a subtype transmitted as a recessive disorder. For dominant gene disorders, a mutation in a nonenzymic protein is expected and may be revealed by molecular genetic approaches that permit the detection of such mutants in DNA (McKusick, 1988).

Three broad classes of models can be described. An introduction to them was provided at the beginning of Chapter 5; now some elaboration will be added to allow preferences to be established. The *distinct heterogeneity* model suggests that schizophrenia is composed of a large number of qualitatively separate disorders, each with its own cause — recessive or dominant genes, trauma, drugs, polygenic systems, bad mothering, etc. —making up a "market basket" of causes with no necessary limitations on causal explanations. Certainly such models have made sense in explaining the transmission of severe mental retardation, blindness, and deafness within families. Within this framework, we might expect phenothiazine-responsive schizophrenia, phenothiazine-resistant schizophrenia, positive-symptom schizophrenia (Harvey and Walker), Type II schizophrenia, paranoid schizophrenia, and so forth, to each be caused by a different locus segregating for its own recessive or dominant genes, or, when nonfamilial, to have a distinctive pattern of neuropathology or pathophysiology. The symptomatic or schizophrenia-like psychoses associated with PCP, crack, and amphetamine abuse can be accommodated by such an umbrella model. The model requires that each posited single-major-locus variety of schizophrenia, whether dominant or recessive, generate appropriate "textbook" pedigrees compatible with such inheritance.

The second model is truly parsimonious, even procrustean, and is called the *monogenic* or *single major locus* (SML) model. It stipulates that all cases of schizophrenia, except the schizophrenia-like psychoses, can be accounted for by one mutated gene at one chromsomal locus. Such a model certainly works for the rare dominant gene that leads to the development of Huntington's disease. Although many researchers have tried to make various versions of this model fit the array of data reviewed in earlier chapters, none has succeeded in accounting for all the important data points. The more feasible versions of the model posit that persons inheriting two doses of the "S" gene—that is, the SS homozygotes—while rare, are always affected, while the SO heterozygotes are buffered from expressing the schizogene to a degree; thus, only 13 percent of the offspring of cases are affected even though 50 percent of them have the putative schizogene. Heston thought at one time (1970) that by extending the analyzed phenotype to include schizoid disease (now called spectrum disorder), a dominant gene model could be fitted without invoking (post hoc) the concept of *incomplete expression*; that is, definitely possessing the gene but not showing symptoms of schizophrenia. Later, this model was seen not to fit a large enough proportion of the family and twin data (Shields, Heston, and Gottesman, 1975). Matt McGue of the University of Minnesota and Gottesman (1989) have shown via the computer simulation of different genetic models for 200,000 families from the general population that families containing more than two schizophrenics in a two-generation nuclear family would be rare; the very rare appearance of pedigrees with more than four cases (only 2 families among 200,000) does not require a single-major-locus model and is quite compatible with a multifactorial-polygenic model.

The groundwork for the third model, the *multifactorial polygenic threshold* (MFPT) model, was laid in Chapter 5 using Figures 8 and 9 (pages 90 and 92). Gottesman and James Shields introduced this model in 1967 for the analysis of data on schizophrenia in an article in the *Proceedings of the National Academy of Sciences*. They had realized that the model devised by Donald Falconer of Edinburgh, Scotland, to explain the genetic transmission of certain nonrare, congenital malformations might be readily adapted to the data available on schizophrenia. They visited him shortly after he had published his ideas in 1965 to discuss their adaptation and received the encouragement they had hoped for.

The hypothetical genotypes involved for all three models described so far are outlined schematically in Figure 17. As a point of departure, a

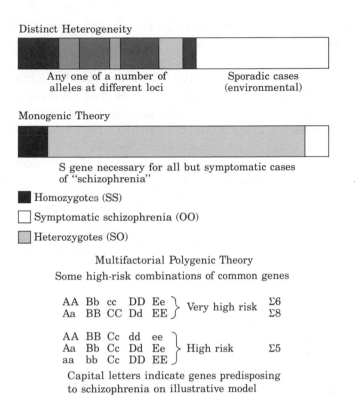

Distinct Heterogeneity

Any one of a number of alleles at different loci

Sporadic cases (environmental)

Monogenic Theory

S gene necessary for all but symptomatic cases of "schizophrenia"

■ Homozygotes (SS)

☐ Symptomatic schizophrenia (OO)

▨ Heterozygotes (SO)

Multifactorial Polygenic Theory
Some high-risk combinations of common genes

AA Bb cc DD Ee } Very high risk Σ6
Aa BB CC Dd EE }   Σ8

AA BB Cc dd ee }
Aa Bb Cc Dd Ee } High risk   Σ5
aa bb Cc DD EE }

Capital letters indicate genes predisposing to schizophrenia on illustrative model

FIGURE 17. Schematic illustration of the three major competing theories to account for the genetic transmission of schizophrenia within families. The total area of the rectangles under distinct heterogeneity, and monogenic theory represent all cases of schizophrenia, and the smaller shaded areas within each rectangle indicate the proportion of cases caused by each illustrated genotype or, unshaded, the environment. (After Gottesman and Shields, 1982.)

relatively simple five-locus version of an MFPT model is illustrated that requires only 2 possible alleles (genes) at each of the 5 loci; this will yield 11 genotypic classes each containing individuals with from 0 to 10 of the hypothesized genes that increase one's liability to developing schizophrenia (see Chapter 5 for a two-locus version). Such genes are indicated by capital letters; the lowercase-letter genes have a neutral effect on liability. This simple version can generate 243 genotypes, but many are repeats of each other; that is, they are functionally equivalent. Some high-risk genotypic combinations are shown, but it is important to note that they are not sufficient for developing schizophrenia — such high-risk combinations must be inserted into one of the epigenetic-developmental models described above in order to determine their outcomes. The hypothetical genotypes determine only the "starting points" at the left side of Figure 16.

Even some very-high-risk genotypes may never become clinically diagnosable as schizophrenics, as evidenced by the discordant identical twins in the Danish studies (see Chapter 6) who, nonetheless, still transmitted their predisposition to schizophrenia to their offspring, in whom it did develop. That some 90 percent of schizophrenics do not have parents with schizophrenia is compatible with this model: Two clinically normal parents could each transmit, say, three capital-letter genes so that their offspring accumulated six, thereby placing them in the very-high-risk genotype category. The phenomenon is just like that seen when two short parents produce a child who grows up to be taller than either one of them, perhaps even crossing the threshold into the National Basketball Association. Within the diathesis-stressor framework, low-risk genotypes — say, those with only three of the hypothesized genes — could, with sufficient liability augmentation from environmental sources and in the absence of sufficient protectors, still develop a schizophrenia, even if mild and remitting. To a large extent, genes and stressors are interchangeable in their effects on combined liability.

It is still premature to reach firm conclusions about which model for genetic transmission is the best. Therefore, it is time to propose an "ecumenical" model or a *combined model.* The identification of individual genes with detectable gene products within a limited-loci polygenic system (wherein just a few loci, say four or five, make a huge difference) is still largely a promissory note from molecular geneticists as we begin this last decade of the twentieth century — but it is a note we expect to be paid off by the start of the twenty-first century. One outstanding example already in hand is the identification of the gene for familial hypercholesterolemia, a defect that leads to abnormal levels of low-density lipoprotein (LDL) cholesterol. It is one gene in a multifactorial polygenic system that is associated with coronary heart disease and a myocardial infarction (heart attack). The gene is remarkably common in the general population — 1 in 500 individuals carries it — but it is not enough by itself to cause heart attacks. In fact, starting the other way round, only some 5 percent of persons experiencing an infarction, have this particular dominant gene; the remainder have arrived at their disease by some other, even more complex, route. Hypertension, another identified vulnerability factor in heart disease, has its own polygenic basis for inheritance. Another route to hypercholesterolemia in the general population is known to be under polygenic control, with a prevalence of 5 percent; many other factors are required over the life span, however, before a subset of per-

Combined Model of Causes

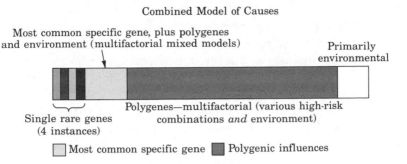

FIGURE 18. A schematic synthesis of an ecumenical, or combined, model showing the most reasonable proportional allocation of causes for schizophrenia after considerations of the major theories plus the hybrid, or mixed, model.

sons with this polygenic inheritance will develop heart disease (Vogel and Motulsky, 1986).

The ecumenical or *combined model*, sketched in Figure 18, permits many flowers (models of genetic transmission) to bloom. One reason that a definitive choice among the three models in the previous figure may not be possible is because each may be correct, but only for a subset of the schizophrenic phenotypes. The combined model as a whole cannot be refuted, a disadvantage to philosophers of science, but it does result in an uneasy peace among warring factions so that they can stop wasting energy fighting with each other and continue to carry on the war against etiological ignorance. Most certainly, disagreements will occur about the amount of explanatory territory to allocate to each submodel in the combined model rectangle.

The proportions shown are not definitive, nor drawn to scale, but represent the personal preferences of the author after digesting the current state of the art of genetic modeling for schizophrenia (McGue and Gottesman, 1989). A minority area is allocated to primarily environmental causes of schizophrenia or schizophrenia-like psychoses for those occasions when they appear to be sufficient to cause the disorder. Another minority area is allocated to the Mendelian genetic varieties, individually rare forms but, in the aggregate, deserving their own area. If, for example, we were creating such a combined model rectangle for mental retardation throughout its range, almost 500 different Mendelian genotypes would fill that area; since each is quite rare, however, they account for only a small proportion of all phenotypes classed as retarded. PKU, for example, accounts for only 3 in 1,000 cases of mental retardation from all causes.

The lion's share of explanatory territory is reserved for two variations of multifactorial polygenic models, both requiring a threshold effect. We would estimate, based on computer simulation studies, that up to one-tenth of schizophrenics within the middle portion of the rectangle could arrive at that outcome by inheriting a specific major gene (similar to the familial hypercholesterolemia gene discussed above) that, together with a number of polygenes plus environmental factors, puts these individuals over the threshold. Meehl's "dominant schizogene" model from Figure 14 is really an instance of a mixed model requiring two different genetic classes of factors plus the environmental contributors. The remaining 90 percent of cases in the middle of the rectangle are there because of polygenic factors plus environmental factors; that is, they reflect the kinds of cases shown at the bottom of Figure 17.

Summary

The following emprical observations garnered from the previous chapters support the preference for some kind of multifactorial polygenic theory for the transmission of the majority of cases of schizophrenia:

1. Environmental models positing factors related to quality of parenting, social stressors, or infectious agents as sufficient causes of schizophrenia do not fit the observed facts.

2. Genetic models positing a single major locus, even with varying degrees of penetrance or expression, do not fit the observed facts.

3. The risk for schizophrenia drops sharply rather than by 50 percent with each step of genetic relatedness as we move from identical twins to siblings and children to grandchildren, nephews, and nieces.

4. The risk for schizophrenia in relatives rises with increasing severity in a proband and with greater numbers of other affected relatives and diminishes with increasing numbers of healthy relatives.

5. Schizophrenia spectrum disorders — diluted versions of schizophrenia —appear in excess in close relatives of probands, even when reared in adoptive homes.

6. Only a minority of pedigrees have other affected relatives; when they occur, the affected relatives may be either on only the paternal (maternal) side *or* on both sides of the pedigree.

7. Possible biological markers of schizophrenia, such as the density of dopamine receptors in the brain, are quantitative traits, not qualitative Mendelian ones.

8. Encouraging analogies exist between the observed facts about schizophrenia and the facts available for conditions with better supported multifactorial polygenic bases, such as coronary heart disease, cleft lip and other congenital malformations, diabetes, and mental retardation.

Conclusions

Information in the preceding chapters converges and can be brought to bear on two major questions: (1) How does schizophrenia develop from its roots to its unwanted fruits? and (2) How is the seed for that fruit transmitted across generations and within families? Although we lack definitive answers, the issues require a systems approach and are best comprehended by a vulnerability/diathesis-stressor model for the first question and a multifactorial-polygenic model for the second question. Present uncertainties are in the process of being resolved by impressive advances in the neurosciences, tempered by a balance between the interaction of reductionism and synthesis, spanning the behavior of molecules and the behavior of human beings.

Indirect and circumstantial evidence for theories about the genesis of schizophrenia has dominated the research field up to now because we have no direct way to measure the hypothetical liability to developing schizophrenia. Liability remains elusive, but we assume it exists so that we can consider models for development and for transmission. Although it is easy to convince reasonable people about the validity of inferring liability in the relatives of schizophrenics, even reasonable people become skeptical about inferring liability in the families of schizophrenics who have no affected relatives, and such schizophrenics constitute the vast majority of cases. We must turn back to the era of Monk Gregor Mendel and, as he did, perhaps accept unseen "genetic factors" as an article of faith, bolstered by our conviction that our models can account for the observations about schizophrenia and patiently wait for advances in the neurosciences.

12

What Next?

The Decade of the Brain

A clue to the energy and excitement surrounding research into the causes and treatments of schizophrenia was provided by the Second International Congress on Schizophrenia Research held in 1989. The more than 250 contributions were divided among the following 10 categories: classification and epidemiology, genetics, neuropsychology, electrophysiology, neuropathology, brain imaging, neurobiology, pharmacology, psychosocial factors, and movement disorders. Despite the diversity of facts presented in the previous 11 chapters, many domains required for a total understanding of the nature and nurture of schizophrenia have been omitted so that we could focus on a few major areas that would tell an important part of the schizophrenia story.

Current literature dealing with research into schizophrenia and mental disorders generally is peppered with such phrases as "on the brink of," "on the verge of," and "on the threshold of." The dictionary tells us that the intention is to convey that we are on the border of "something towards which there is progress," and that "something" in this case is significantly incremental knowledge about the nature and nurture of schizophrenia. Such phrases are not mere hyperbole. Contemplate the following facts: 95 percent of what is known about brain function has been acquired in the past 10 years; there is a 10- to 15-year lag between basic science advances and their application to practical problems of illness; the interdisciplinary Society for Neuroscience has increased its membership from 250 in 1971 to more than 11,000 today; in the past 5 years, the number of articles with the word *brain* in their titles has doubled to 100,000 publications indexed in the National Library of Medicine; and tools and techniques that permit the observation of brain structure and function in living patients have only begun to be used in a few high-technology centers (see Early, Hari, Pardo, Posner, and Suddath citations).

No wonder, then, that the National Advisory Mental Health Council (NAMHC) to the NIMH has recommended to it and to Congress (Public Law 101-58, July 25, 1989) that the Decade of the Brain be initiated for the 10 years leading us into the year 2000 and the start of the twenty-first century. Among the innovative, essential, and costly initiatives being requested are the following: a National Neural Circuitry data base; a Computer Applications in Neuroscience program; a Molecular Neurobiology program; Centers for Neuroscience in Mental Illness; a National Brain Bank program (two are currently in operation in the United States, one in the United Kingdom); a National Tissue Bank (to get cells for DNA studies); and adding Developmental Neurobiology to the NIMH Intramural Research Program. Congress has already authorized the gigantic Human Genome Mapping Project under the direction of James B. Watson, codiscoverer with Francis Crick of the structure of DNA. Resistance to the costs involved should be overcome by the simple observation that the continuing direct costs for clinical care of the victims of psychiatric disorders alone are more than $40 *billion* per year and the continuing indirect costs from lost wages and productivity are more than $70 *billion* per year. Although an annual research budget for neurosciences at the NIMH of $30 *million* sounds like a generous amount, it amounts to only 75 *cents* per person experiencing diagnosable mental disorders per year.

As the NAMHC (1988) concluded, "Given the funding of such initiatives, society could look forward to a detailed map of specific gene products, specific transmitter molecules and their receptors, specific cell types, specific cell circuits, and specific behavioral functions in the primate brain, with direct extrapolations testable in the human" (p. 7).

The Council also promulgated a more schizophrenia-specific white paper with contributions from seven panels of experts on the whole gamut of elements important to schizophrenia. Their National Plan for Schizophrenia Research has been published in the *Schizophrenia Bulletin* (1988, No. 3), providing an excellent bibliography on recent developments following each of the panel recommendations.

Neurobiology and Schizophrenia

Only the "flavor" of the exciting advances in the neurosciences relevant to schizophrenia can be provided in summarizing statements, but the interested reader can pursue in-depth treatments via the bibliography for this chapter. "How does the brain really work?" is no longer a rhetorical question. The accumulation of partial answers to how it works, as well as how it malfunctions, feed directly into questions about the causes and treatments of schizophrenia (and other forms of psychopathology). An adult brain contains some 100 billion neurons and the typical neuron has about 1,500 synapses, or connections, to other neurons; the total number of synapses across which electrical/chemical signals are transmitted or blocked is mind-boggling. At each synapse, there may be a million receptor molecules that mediate the communication with other neurons, families of chemicals known collectively as neurotransmitters and neuromodulators.

The king of putative neurotransmitter dysregulation in schizophrenia is dopamine, but there are other contenders that cannot yet be ruled out, such as serotonin, acetylcholine, and neuropeptides (see Meltzer, 1987; Friedhoff et al., 1988). Driving the dopamine hypothesis about the "intermediate cause" of schizophrenia is the idea of *overactivity* of neurons containing or using dopamine in mesolimbic, basal ganglia-nigrostriatal, and mesocortical regions of the brain. Because the most effective drugs used to treat schizophrenia block dopamine receptors (hence, transmission of signals), and because agents that make schizophrenia worse — such as PCP, crack, and amphetamines — increase dopamine levels, the hypothesis has made sense and has had remarkable survival in a rapidly changing bioscience environment.

P. J. McKenna (1987) of the University of Leeds, emulating Sherlock Holmes at his finest, traces the clues throughout the various dopamine-dependent circuits of the brain to tie together symptoms and behaviors most impressively. He can even explain the mystery of auditory hallucinations as possibly being the consequence of a failure of dopamine circuits in the anterior cingulate cortex that may control a "subtractive" function of auditory decoding that ordinarily permits primates to discriminate between their own ongoing or intended vocalizations and external auditory information.

Such indirect evidence about the role of dopamine is now being replaced by still controversial direct evidence obtained from the use of PET to visualize the density of D2 (one specific kind of dopamine receptor) receptors in the caudate-putamen regions of the brains of living schizophrenics (Buchsbaum and Haier, 1987; Wagner et al., 1988; Waddington, 1989). Earlier studies using the brains obtained at autopsy from a small number of schizophrenics had suggested that there would be an increase of dopamine receptor density, but such findings could be attributed to an "adaptive" response of the brain to being treated with neuroleptic drugs that blocked dopamine transmission.

Dean Wong, Henry Wagner, and their nuclear medicine colleagues at Johns Hopkins University were able to find so-called drug-virgins among young schizophrenics — that is, patients who had not yet been treated with neuroleptic drugs — and examine them with the PET using special radioactively labeled chemicals (ligands) to mark the D2 receptors in the brain. Strikingly elevated levels of the D2 receptors were found, thereby strengthening the postmortem findings and the general dopamine hypothesis. However, another highly regarded group of scientists at the Karolinska Institute in Stockholm, Sweden, found *no* elevations of the D2 receptors in their sample of drug-naive schizophrenics — an unequivocally negative report. Sedvall, Farde, and their Swedish colleagues used a different ligand, a different intricate mathematical model for making inferences, younger patients, but the same DSM-III criteria.

The Wong group (1989; Tune et al., 1989) has replicated its results with an enlarged sample and has extended the D2 brain receptor findings to drug-naive patients with bipolar affective disorders. The surprising result now is that the expected specificity for schizophrenia has been eroded; psychotic bipolar cases also have markedly elevated numbers of D2 receptors, perhaps implicating this factor as a general one for disorders marked by delusions and hallucinations.

PET research with psychiatric patients is still a rare bird, and much more remains to be learned; very few of these expensive scanners are available, and the research requires a cyclotron as well as a nuclear chemist to prepare the ligands in timely fashion to interface with the availability of patients. PET has been used more often to measure the amount of work done by different areas of the brain, as indexed by the brain's use of radiolabeled glucose (but see Early et al., 1989 and Posner et al., 1988). For the first time, such a technique has been used by

Susan Resnick, Raquel Gur, and their colleagues at the Brain-Behavior Laboratory of the University of Pennsylvania with the identical twins discordant for schizophrenia from the Torrey-Gottesman and colleagues project at Saint Elizabeths Hospital (see Chapter 6). In initial work with five pairs and the ill twin remaining on neuroleptic medication, the metabolism of glucose was elevated in the brain region called the basal ganglia for each schizophrenic twin; it was not elevated in any of the healthy cotwins. Time will tell whether the evidence for increased subcortical metabolism, in accordance with other nontwin findings, can be attributed to antipsychotic medication only. PET has the potential to revolutionize our understanding of mental disorders and their effective treatments.

In a masterful, integrative review modestly entitled "Observations on the Brain in Schizophrenia," Daniel Weinberger and Joel Kleinman, section chiefs at Saint Elizabeths Hospital/NIMH, separate wheat from chaff in classical and contemporary high-technology research. On structural changes in the brain found at autopsy, they conclude that no consistent changes have been identified. "Nevertheless, evidence of nonspecific degenerative pathology, especially in the limbic forebrain (for example, amygdala, substantia innominata, medial pallidum, and hippocampus) has been described in controlled studies by several investigators. Whether the findings are causative, coincidental, or consequent to the illness is not known" (p. 55). On postmortem changes in neurochemistry, they conclude that while no consistent findings have emerged, a consistent *trend* appears. They note that no postmortem data are available from the simultaneous study of both anatomy and chemistry, a disconcerting gap. "The vast majority of the neurochemical findings involve the limbic system and the striatum and, to a lesser extent, the frontal cortex. . . . The next question is whether the changes reflect pathology intrinsic to these regions, or pathology intrinsic to other brain areas involved in the neurochemical regulation of the limbic system and striatum [producing secondary changes]" (p. 60).

Overall, Weinberg and Kleinman conclude that the evidence for a definitive brain lesion in schizophrenia "is no more convincing now than it was at the turn of the 20th century. . . . Whether all the clinical manifestations of schizophrenia will ultimately be referrable to pathology in one site . . . or be the result of a more generalized . . . process, is an issue for continuing investigation. . . . In conclusion, it can be stated that evidence linking schizophrenia to organic pathology of the brain is on the verge of being definitive" (p. 62). I, for one, am inclined to believe them.

Few authors address the difficult issue raised in the previous chapter of how to reconcile structural and functional changes in the brain with the episodic course of schizophrenia from dysfunction to social remission and back again to dysfunction. In an imaginative tour de force dealing with neural plasticity in schizophrenia, John Haracz of the UCLA Brain Research Institute draws together the principles of genetic variation with the observations on neurobiology to speculate that behaviorally important changes in neuronal connectivity might explain the phenomenon and be produced by "genetic variations in at least six parameters: (1) the migration of neurons, (2) the distribution of neural pathways, (3) the number of neurons, (4) the orientation of dendrites, (5) the sensitivity of receptors, and (6) the plasticity of neuronal connections in response to environmental changes" (p. 206).

In a remarkable finding from the developmental neurobiology of the canary's brain, it has been determined that even in adult birds *new* neurons can be produced in response to song, doubling in one region within seven weeks; these extra neurons died when the male canary stopped singing after the mating season, new neurons taking their place on a seasonal basis. Is it too "birdbrained" an idea to suggest that such plasticity in genetically predisposed persons could fill in the speculative gaps in the "boxologies" for the epigenesis of schizophrenia sketched in the previous chapter?

The Continuing Promise of Longitudinal High-Risk Studies

By careful examination of schizophrenics *before* they become dysfunctional, it should become possible to distinguish between those facets of disturbance that are indicators of the genetic predisposition to developing schizophrenia and those that are signs of incipient schizophrenia or that may be accommodations to the disorder itself. Efforts at retrospective reconstruction of the development and origins of mental disorder are notoriously unreliable. An efficient strategy for accomplishing this difficult prospective task was suggested by John Pearson and Irene Kley of the Mayo Clinic in Rochester, Minnesota, in 1957. Pearson and Kley were mainly interested in designing prospective longitudinal high-risk studies of Huntington's disease, with a base rate of 1 in 20,000, but also illustrated their thinking with the example of schizophrenia. They reasoned that rather than study 1,000 members of the general population for half a century to get a yield of 10 schizophrenics (the projection from the usual risk of 1 percent), one could study 100 babies or children born to schizophrenic mothers and get a yield of about 13, with some cases appearing after *only* a 15-year wait (see Chapter 5). Thus was the "cost-effective" high-risk strategy for schizophrenia research conceived. The detailed histories and interim results of such heroic projects may be found in the book *Children at Risk for Schizophrenia — A Longitudinal Perspective*, edited by Norman Watt, E. James Anthony, Lyman Wynne, and Jon Rolf.

The most intensive and long-lived of any of these projects was begun by Barbara Fish, a child psychiatrist at UCLA, while she was at New York Bellevue Hospital in 1952. She began with babies born to 2 schizophrenic mothers and later added 10 more; she continues to follow them, wherever they are, in 1990. Her concept of *pandysmaturation* merges with Meehl's notions about neurointegrative defect and schizotaxia; Fish believes she has data to support the idea that schizophrenia can be detected as early as infancy by careful attention to irregularities in cognitive, neurological, physical, and personological growth. Her labors will soon come to fruition (1987).

When Sarnoff Mednick, an American psychologist, and Fini Schulsinger, a Danish psychiatrist, began to collect data for their high-risk project in 1962, little did they think that they would still be at it today (Mednick, Cannon, Parnas, and Schulsinger, 1989). Their 27-year follow-up of 207 children examined in adolescence, born to schizophrenic mothers and their controls, continues to excite a wide audience. Fifteen of the now-grown children have developed schizophrenia over the course of being observed and tested, and the investigators expect another 15 before they are through. A very wide range of variables have been studied, from psychophysiology to MMPIs (Minnesota Multiphasic Personality Inventories) to CT scans. Definitive results are still in the offing.

Another huge and productive undertaking was launched in New York City by Nikki Erlenmeyer-Kimling and her colleagues in 1971 by enrolling the first 355 children between the ages of 7 and 12 years who had been born to schizophrenic mothers or fathers (124), to other hospitalized psychiatric controls (65), or to normal controls (166). Her program emphasizes genetic theories and measures of attention and information processing. At the Dahlem Conference in Berlin (1987), Erlenmeyer-Kimling reviewed biological markers for the liability to schizophrenia from the vantage point of high-risk studies. She concluded that eye-tracking dysfunction and deficiencies in sustained attention looked promising as such markers; that electrodermal activity — e.g., GSR (galvanic skin response) and ERP (event-related potentials) — and platelet monoamine oxidase activity were no longer interesting; and that not enough consistent data were available for neurological signs, ventricular brain enlargement, and interhemispheric integration to reach a conclusion. In her own sample, as well as in the one being studied at UCLA by Keith Nuechterlein, measures of sustained attention derived from complex versions of the Continuous Performance Test (previously developed to assess brain injury) looked very promising as biological markers, with the children of schizophrenics performing poorly in discrimination tasks (27 percent vs. 11 percent vs. 6 percent in the Erlenmeyer-Kimling groups). One facet of the New York program also suggests that the personalities (measured with the MMPI) of a subgroup of high-risk offspring are conspicuously different from those of the other high-risk children in adolescence and also are different from those of the two control groups (Moldin, Gottesman, and Erlenmeyer-Kimling, 1990). Definitive results are yet to be obtained.

Valid childhood predictors of adult schizophrenia may indeed be forthcoming from high-risk studies of the young offspring of schizophrenic mothers and fathers, but three possibilities (Hanson, Gottesman, and Heston, 1990) must be considered before such predictors are equated with biological or possibly even genetic markers for schizophrenia: (1) the deviance may merely reflect maladaptive reactions to being reared by and with a parent with major mental illness (hence, the importance of complementary prospective adoption study designs); (2) the deviance may reflect nonspecific liability to developing schizophrenia — that is, potentiators or correlates of potentiators for the specific liability itself (hence, the importance of *psychiatric* as well as *normal* control groups); and (3) childhood predictors might represent the earliest signals of already started schizophrenia (hence, not useful for identifying contributors to the liability, but very important as an early warning signal for the initiation of intervention). A fourth possibility is that the group of "outliers" having deviance in childhood arose by chance and small-sample fluctuations; only 4 of the 15 schizophrenic "hits" in the Danish high-risk study had been identified at an earlier age as being the ones who would go on to develop schizophrenia, and 20 others (false positives) had been selected at an average age of 15 as the unlucky ones. The proof of the pudding in high-risk research designs will come only from thorough follow-ups of the children until the expected yield of bona fide schizophrenics emerges, unless valid genetic markers are discovered in the meantime.

Genetic Markers and Genetic Linkage

The current breathtaking pace of discovery of genes affecting health is so phenomenal that it can be followed on the front pages of the *New York Times* and the *Wall Street Journal* and on the evening television news without even reading such scientific journals as *Nature*, published only weekly. Such breakthroughs have virtually all been relevant to major gene disorders; that is, the rare Mendelian forms of disease with their prototypical 50 percent and 25 percent recurrence risks among siblings. It is only a question of time and informed initiative before the revolution in human genetics ushered in by David Botstein, Ray White, and their colleagues in 1980 reaches the more complex phenotypes of the major mental disorders such as schizophrenia. Botstein and colleagues showed how to construct a refined genetic linkage map of homo sapiens using the newly discovered restriction fragment length polymorphisms (RFLPs), formed by cutting up DNA derived from ordinary blood samples with special enzymes known as *restriction endonucleases*. Until RFLPs came along, the possibility of actually proving that a putative familial disease gene was physically very close to a known gene on the same chromosome —*genetic linkage*—by the analysis of the pattern of disease in large pedigrees, was greatly handicapped by the availability of only 30 to 40 classical markers such as the ABO blood-group genes and the HLA antigens.

Just as the road from New York to San Francisco is too long to be signposted adequately by 40 markers, so too is the human genome too long to be mapped by only 40 genetic markers. Since more than 3,000 RFLPs have been added to the classical markers, however, it has become feasible to discover linkage for Mendelian traits and to lay the groundwork for establishing linkage for more complex traits such as schizophrenia and affective psychoses. In 1983, James Gusella, Nancy Wexler, and their colleagues reported the discovery of a marker gene for Huntington's disease on chromosome 4 based on very informative pedigrees from Venezuela; other discoveries have followed in regard to cystic fibrosis, muscular dystrophy, and other Mendelian diseases.

In a modern classic paper entitled "Genetic Diversity, Genome Organization, and Investigation of the Etiology of Psychiatric Diseases," Robert Cloninger, Theodore Reich, and Shozo Yokoyama of Washington University in St. Louis provide a partial blueprint for the next decade of needed research into the origins of mental disorders. They make the crucial point that even though there may be between 50,000 and 130,000 structural genes (such genes make proteins; regulatory genes turn them on and off) in our species, genome organization involves so much backup redundancy that *only* 3,000 to 9,000 functional proteins are produced by gene clusters. Clearly, the tasks of the next decade are manageable. Cloninger et al. conclude, "A complex phenotype cannot be understood by information on DNA sequences only. Rather, we must unravel the etiology of complex traits by analyses beginning with phenotypic heterogeneity and then progressing to study heterogeneity in underlying biosocial risk factors. . . . Accordingly research should consider multiple biosocial risk factors that mark different steps in the genotype-phenotype pathway, some sensitive ('distal' markers) and others specific ('proximal' markers)" (1983, p. 243). With one exception, the search for genetic linkage to schizophrenia with the available genetic markers has not been productive (Gurling, 1986; McGuffin and Sturt, 1986). The exception, to be described shortly, has not been confirmed. Anne Bassett, then (1987) a young psychiatrist at the University of British Columbia, was so well tutored that when she admitted a 20-year-old male college student who met criteria for a DSM-III diagnosis of undifferentiated schizophrenia, but who also had subtle dysmorphic features of the face and head (jutting forehead, widely spaced eyes, folded-over, "stuck-on" ears), she seized on his mother's casual comment that the boy resembled his *uncle who also had become affected with schizophrenia at the age of 20.*

Brilliant detective work quickly led to the discovery, upon cytogenetic examination, that the chromosomes of the uncle-nephew pair had the very same abnormal karyotype — an unbalanced translocation whereby extra material from chromosome 5 had been inserted into chromosome 1 shortly after the zygote (fertilized ovum) was formed. Both schizophrenic males have a normal pair of chromosome 5s, as well as an additional copy of a region of chromosome 5 that high technology can determine to consist of a segment on the long arm of chromosome 5 identified as "5q11.2 to 5q13.3" that now exists where it does not naturally belong, in chromosome 1. The men are said to be *trisomic* for that segment. The $64,000 question becomes, "Is there a gene or genes in that segment which, in triplicate, is related to the presence of schizophrenia?" All geneticists are mindful of the fact that triplication or trisomy of the genetic material on chromosome 21 leads to Down's syndrome.

But the story goes on. The mother, who is psychologically within normal limits and has no physical findings of note, has the very same abnormality of her chromosome 1 as does her son, that inserted segment from her chromosome 5, *but* she has a missing part, or deletion, of that segment from her chromosome 5, leaving her with a *balanced* translocation; that is, no *extra* parts. Thus, the triplication of 5q11.2 to 5q13.3 must be the culprit in this family. Furthermore, William Iacono at the University of Minnesota and colleagues (1988) have examined eye tracking (an indicator of attentional processes) in the two schizophrenics, the mother, and two other normal relatives; only the schizophrenics have an abnormal eye-tracking record.

Between the time Bassett reported her initial observations at a scientific meeting and the time of publication, the findings were influencing other groups of investigators to focus on chromosome 5 using various RFLP probes. If the clue panned out, it would be a tremendous time-saver, because all the other chromosomes would not need to be examined for linkage markers — a very lengthy procedure, to say the least. (It was only luck that led the Huntington's disease team researchers to start their probing for linkage on chromosome 4.) A big media and science splash was created when, in November 1988, Hugh Gurling, Robin Sherrington, and their other team members in England, Iceland, and California, reported in *Nature* what looked like strong evidence for a dominant schizophrenia-susceptibility gene on chromosome 5 localized to a region overlapping with the region implicated by Bassett and colleagues. Gurling's team made use of RFLPs from chromosome 5 as linkage markers to track the pattern of schizophrenia and other personality disorders in five Icelandic and two British pedigrees with 104 members, loaded with mental illness — 39 cases of various subtypes of schizophrenia, 5 schizoid personalities, and 10 "fringe phenotypes" (a mishmash of pathology, very common in the general population). Depending on the definition of a "hit" for a schizophrenia-related illness and certain assumptions about the penetrance of the putative gene, Gurling's team obtained *lod (logarithm of the odds) scores* ranging from 3.2 to 6.5 as evidence of linkage between the disease and the markers on chromosome 5. A lod score of 3 implies odds of 1,000 to 1 against it happening by chance and a score of 6 implies odds of 1 million to 1 against a chance finding.

On the same page of *Nature* on which their article ended, a new one by Kenneth Kidd, James Kennedy, and their team members at Yale University, Stanford, and the Karolinska Institute in Stockholm found *strong evidence excluding linkage* for some of the same chromosome 5 RFLPs as used by Gurling. Kidd's team used heavily loaded families from the Northern Swedish isolate studied years ago by Jan Böök (1953) and recontacted for interviews and blood samples to extract DNA. Eighty-one subjects from seven branches of one large pedigree were examined; 31 met DSM-III criteria for schizophrenia. Böök's schizophrenics have been noted in the past to differ clinically by being mostly catatonic varieties and, more recently (Kennedy et al., 1989), by being neuroleptic nonresponders.

Certainly, etiological heterogeneity, our old nemesis, could explain such divergent findings between the Swedish and Icelandic-British pedigrees. Neil Risch, a brilliant theoretical population geneticist at Yale, notes the irony (1990) in calling nonreplication of results, or failure to confirm them, evidence of genetic causation in terms of a second, independent, never-observed major locus. Surveying the rash of reports identifying single major genes for manic-depressive disorders in Amish and Israeli pedigrees (findings now in limbo) on chromosomes 11 and X, respectively, and for Alzheimer's disease on chromosome 21, *none* of which had been replicated, Risch comments on "the remarkable consistency of such inconsistent findings in the recent history of linkage analysis of common neuropsychiatric disorders" (p. 4). Like the present author, he prefers a disorder such as diabetes, a complex phenotype, as a model for the common mental disorders, reserving single-major-locus explanations to the place reserved for them in Figure 18 (page 231).

L. Leigh Field (1988), a medical geneticist at the University of Calgary, calls insulin-dependent diabetes a model for the study of multifactorial disorders. Yet it was not until 1986 that a consensus of diabetes researchers could be reached that all simple Mendelian models had to be relinquished (Baur, 1986). Viruses, diet, lifestyle, maternal effects, and *some* genes of large effect in the HLA region of chromosome 6 from a larger "system of genes" throughout the genome have all been implicated in a multifactorial system related to diabetes.

In the meantime, Bassett, Conrad Gilliam, Charles Kaufmann, and colleagues at the New York State Psychiatric Institute and Columbia University have pulled off the remarkable feat of cloning the chromosome 5 with the deleted segment from the unaffected mother of the Canadian schizophrenic (with fully informed consent) and now have an unlimited supply of a line of cells formed by fusing human with hamster cells. DNA markers can now be mapped to show whether they are out of, or in, the triplicated segment. A number of other research teams are in the process of reporting no linkage in this region of chromosome 5 (McGuffin et al., 1990).

Bassett (1989) offers a balanced perspective on the entire linkage enterprise and begins to anticipate the ethical issues that will arise in the event such efforts are successful. Because the gene or genes may never be expressed (see Chapter 6), how would information about genotype be conveyed? Would prenatal genotyping be grounds for abortion? Who really owns the cell lines, given that they may be useful commercially in creating treatments for ameliorating or preventing schizophrenia? Risch and Bassett both caution that genetic linkage analysis, one of the glitziest techniques in bioscience (cf. MEG — magnetoencephalography — PET, SPECT — single photon emission computed tomography — rCBF), is beset by its own methodological limitations and requires circumspection before calling a press conference.

Given the message in Figure 18 about the multiplicity of causes and their relative weighting in a combined-ecumenical model, the eventual demonstration of major gene varieties of schizophrenia can be seen in perspective. Take Huntington's disease as a genocopy of schizophrenia —20 percent of cases in the literature are misdiagnosed on presentation of symptoms that overlap with those of schizophrenia—and ask, What proportion of schizophrenia-like psychoses are caused by the mutated dominant gene on chromosome 4 that leads to HD? By dividing the two population base rates for HD and schizophrenia, 5/100,000 by 139/10,000 (see p. 74 for this close to 1 percent value) and taking the 1/5 who are misdiagnosed, we obtain a value of 7 in 10,000 schizophrenics who have a single major gene for schizophrenia-like psychosis; that gene would have been identified by studying enough specially loaded pedigrees with RFLP's *and* the lucky choice of chromosome 4 as a starting place. The remaining 9,993 cases of schizophrenia would have other causes. This is not to belittle the major gene variety that could be identified, because it might provide direct information about gene products (called reverse genetics) that could determine the mechanisms involved in the final common pathway toward schizophrenia, thereby laying the groundwork for the prevention and amelioration of this devastating disorder.

Conclusions

Although formidable problems still remain for the clinicians and researchers who deal with schizophrenia as a treatable and preventable human condition, as well as for the patients and their families and society as a whole, we appear to be on the verge of resolving those problems as we start the Decade of the Brain. The pace of discoveries in the neurosciences and the beginnings of synthesis and integration of disparate results from interdependent approaches by scientists with growing sophistication encourage such optimism.

Now that the genie is out of the bottle for rare genetic disorders, the more common and complex ones such as schizophrenia, major affective disorders, and Alzheimer's disease/senile dementia will yield their secrets to advances in brain imaging and in molecular genetics, aided and abetted by refinements in description and diagnosis and the models of biomathematicians. Once again we call on the perspectives of Thomas Huxley to remind us that tragedy in science is "the slaying of a beautiful hypothesis by an ugly fact." The players in, and the victims of, the schizophrenia story welcome this kind of tragedy because it quickly gives rise to newer, potentially viable hypotheses demanding new "ugly facts." The reward for enduring the transient "agony of defeat" is the "ecstasy of victory," for with it will come the ability to untwist the twisted molecules and thence to untwist the twisted minds.

A Story to End the Story

Giving Love . . . and Schizophrenia

A mother who is herself schizophrenic reveals her self-doubt as a mother to her young child and her fears about transmitting her illness to him.[1]

Five years ago, I looked forward with concern to speaking to my 3-year-old son's nursery school teacher. I was finally recuperating from a chemical imbalance that had caused a complete psychotic breakdown, four hospitalizations over an 8-month period, and grouped me with the two million other Americans who suffer from the schizophrenias.

Since I was becoming physically able to get out more, I had just attended a lecture on learning disabilities. The therapist's final comment stayed on my mind: "If a child is not speaking in simple sentences by the age of 3 and cannot act on several consecutive instructions (such as: 'Please go upstairs, get my handbag, bring it down here, and put it on the chair.'), it is a sign of possible difficulties in school." Ivan wasn't speaking in sentences, nor could he usually follow one instruction — let alone three. Also, after my husband Terrance picked up our son at school, I asked him if he had peeked through the small pane in the classroom door, as I enjoyed doing, to catch a glimpse of Ivan playing with the other children. He had, but all the other children had been in a circle with the teacher while Ivan was off alone in another area playing with the trucks.

I was concerned. In answer to my question of how Ivan was doing, his teacher replied, "He's slow." Then she quickly qualified it. He didn't appear to be mentally slow, but he was always the last to finish a project,

[1]M. DuVal, *Schizophrenia Bulletin,* 1979, 5: 631–636.

to line up, or to put away his toys. Sometimes she even had to remove him bodily from whatever he was doing.

At a more formal conference the litany continued. His speech was far behind that of other 3-year-olds. He lacked confidence in his own abilities. He was aggressive at times, withdrawn at others. He was moody; a reprimand could cause him to burst into tears. His behavior was unusual: When another class had come in to join in a songfest, Ivan had removed himself from the circle and sat, of all places, in back of the teacher. His approach to his work was indifferent. Was he stubborn, or was he lazy? He was a difficult child to interpret and understand.

It was the contradictions in his behavior and his inappropriate behavior (such as sitting behind the teacher instead of with the group) that beamed the warning light of possible schizophrenia. The extremes in his behavior and his tuning out of the world so that the teacher had to sometimes remove him bodily reinforced my concerns.

Shock treatments [ECT or electroconvulsive therapy] . . . had controlled my illness [very infrequently used for schizophrenia nowadays, but may help with coexisting *severe* depressive symptoms or uncontrolled, catatonic excitement or stupor] so that I no longer had to take the psychiatric drugs (called "Downers" on the street) that had placed me in the twilight world of the mentally ill. I felt like myself again, and that night I sat at the kitchen table to watch him objectively, satisfied that no longer would I have to exert all my control not to let my child see me nodding like an addict or enduring the involuntary jerking [tardive dyskinesia] of my body from the dangerous potency of the drugs.

But it was the beginning of 1975, and for almost a year I had sat at the kitchen table in the evenings too sick to do more than pull all of my energy together for one or two brief excursions into play with him and then leaving him to his own resources. My illness had also come during the energy crisis of 1973–1974, and although the worst of the crisis was over, Terrance and I were still practicing fuel and power conservation by keeping almost all our lights off and our thermostat low.

As I watched, Ivan played silently in a corner with his cars and trucks. Terrance was usually away working in the evenings, and tonight was no exception. I suddenly realized that almost from the time Ivan was born, because of his father's hours and because he was an only child, he had lost the ideal adult chitchat or the prattle of other children that goes on hour after hour in a family that encourages the development of speech patterns. The gas shortage had kept visitors to a minimum, my illness had occurred during the traditionally isolating winter months, it had been difficult for me to drive under the influence of the drugs, and there were no children his age on the block. His frequent playmate before my illness had been a partially deaf 3-year-old who was also not talking.

Ivan had started verbalizing early — 8 months — but when my illness occurred, his speech and his personality regressed sharply: "Tater chips" (potato chips) had become "T-B-B." His favorite game was "Bebee." He would "cry" like a baby and then wait to be cuddled on my lap. When it was time to watch his favorite children's show, I knew he would crawl up on my lap to see the show from the security of my body.

Life had become insecure for Ivan since that day I had been wrenched from him. My parents, in-laws, sister, brothers, and friends were touching in the eagerness to be of service. My mother came in to help, then my mother-in-law, and then Ivan spent time with his godparents. Finally, Ivan's pediatrician stepped in when he noted Ivan's unhappiness during an office visit and recommended that he stay with one person while I was hospitalized, preferably someone with young children, even if it meant boarding him out — which eventually happened. Ivan's godparents found such a woman who has my lifelong gratitude for truly treating Ivan as if he were her own child.

My appreciation for this woman knew no bounds when I looked for a woman closer to home to care for Ivan 1 day a week while I recuperated. Any mother, sick or well, my female psychiatrist has counseled, needs at least 1 day a week away from a 2-year-old. The woman I found gave excellent physical care (typical breakfasts were bacon and eggs or waffles) and didn't dump the children in front of a television set, but there was little personal warmth that I could see. When I picked up Ivan at the end of the day, all the children would be in her entrance-way sitting in their miniature chairs, forming a razor-sharp line as they clutched their individual diapers and clothing bags. A least, I consoled myself, Ivan was having the experience of other children and the day of relaxation was a help. It was also an advantage having my mother close enough to take him on extremely short notice on the days when I was just too ill to care for him.

Terrance's help came in the few hours he was at home. He had been a typical new father — proud of his son, eager to diaper him and later take him fishing, to feed the ducks, and to make him his assistant as they puttered with tool boxes and ladders — comrades in chaos. When my illness hit, he doubled his already extensive efforts at his job to pay for the overwhelming cost (even with insurance) of medical and pharmaceutical care, and for board and child care for Ivan. Terrance's first concern was having me get well by providing the best skill available. His next concern was juggling his job with running a household and coping with the increased psychological responsibility of Ivan. After the first few days of the kind of life many women by choice or necessity lead today, he confessed to me in the hospital that he never knew I went through so much in the course of a day. His attitude was tinged with wonder at the endurance level of women. . . .

As he stayed in his corner, silently playing, no expression on his face — perhaps waiting for the other extreme of my suddenly rousing myself to play with him vigorously or to read to him, the one light outlined him cruelly as I absorbed my realization. Without knowing it, I had brought him to this turning inward — to this becoming almost totally unaware of his outside world. In giving him my love and trying to carry on through the haze of my illness, it appeared to me that I had also been giving him the early symptoms of schizophrenia.

I began to think of all I had read and learned about my illness. . . .

Although I knew from my reading that Ivan stood a 10 percent chance [too precise; see Chapter 5 for a range from 2 percent on up, depending on pedigree] of having inherited enough genes to become schizophrenic, I did not believe he was actually ill with the disease. His present behavior appeared to be a defense against the constant traumas of being separated from and then reunited with his mother.

I remembered when a friend from another city came to visit after my first hospitalization, and her horror upon my return from a quick trip to the store two short blocks away. In Ivan's anguish at my leaving, she related in stunned disbelief, he had almost torn his hair. After enduring the constant pain of sudden separations, Ivan eventually seemed to have decided that it would be less painful if he restrained his capacity for awareness, if he retreated more and more into his own corner with his cars and trucks.

Although I did not believe that Ivan was ill with schizophrenia, I did believe that just as he had inherited other genetic characteristics from me, his having inherited a predisposition for schizophrenia was a possibility indeed. And, just as he had absorbed aspects of my healthy personality, he had also absorbed aspects of my ill one. Add to that the adverse, stressful conditions under which he had to live, and I thought that if he were inclined toward the schizophrenias, whatever unknowns hold the schizophrenia at bay could be eroded. [This paragraph is a remarkably clear formulation by a layperson who happens to suffer from schizophrenia.]

Since my many minutes of detached observation had made it frightfully clear that his environment was the other culprit, I got up from the table and, disdaining the energy shortage, turned on all the lights on the first floor to open up and brighten his world. I moved up the thermostat to a more comfortable 73 degrees. I turned on the radio to a talk show so he would gain a sense of the continuity of speech. (I thought of my past extremes: I was either playing with him, or there was absolute silence.)

To work on his senses, I invented a game called "What Do You See?" which we started playing that very night. I placed my hands over his and his own hands over his eyes, and we marched soldier-fashion around the first floor stopping at various objects. I would take his hands away from his eyes, gently slap his hand on the object and ask, "What do you see?" In this way I began to build up his vocabulary using household items as my tools. I felt ashamed that words like "refrigerator," "counter," and "coffee pot" were terms that appeared to be new to him.

A take-off on "What Do You See?" that I learned about from his teacher was "What Do You Smell?" I would blindfold him and then put various items such as cake, coffee, and soap in bowls. He would sniff each bowl and tell me in a complete sentence (I would only accept a complete sentence) what each bowl contained. We had always baked together a great deal so measuring, breaking eggs, stirring, whipping, beating, and the achievement of fudge brownies or a mile-high cake continue to be a big thrill for him.

Since I knew that in schizophrenia there is often a feeling of unreality, of floating, or of not being sure where one's body begins and ends, I invented another game to ground him in reality—to make him feel himself as a total human being. I would shout "chin" and then nibble his chin with my lips. Then "ear," nibble an ear, and so on through all parts of his anatomy. He would shriek with glee with each nibble, and through his laughter was also feeling where he began and ended. At the same time he was adding parts of the body to his vocabulary.

When it was time for bed that night, I instructed Ivan to begin taking off his clothes. He looked at me perplexedly and whimpered, "Why?" I realized that I had been dressing and undressing him because it had been easier for me to do so. It's easier to care for a baby than contend with a toddler. I had been treating him like a baby, not leading him step by step up to and past the increasing responsibilities of a developing child. The reasons were deeper than the apparent not being willing to let go.

Terrance and I had waited 5 years for our son. When he arrived, it was a delight for me to be at home with him. I cuddled and interacted with him endlessly. For both Terrance and me, it was like playing house with a live doll. When we found we were going to have another baby we were ecstatic.

Our second child miscarried at 3-$\frac{1}{2}$ months. He would live for 3 hours as I lay in my hospital bed, waiting for the final anguish of his death. Of course, there was no hope that he would survive. One thing that sustained me was that I had a "baby" to return home to and recuperate for. The "baby" was then 1-$\frac{1}{2}$ years old.

When our second child was due to have been born, I began to experience some of the physical symptoms of childbirth such as lactation and also the euphoria many women have after giving birth. The day after our child's scheduled birth, because of biochemical abnormalities, I became completely psychotic with schizophrenia.

I took a picture of Ivan with me to the hospital. He was a bubbly 2 months old in the picture, and I kept it on my headboard, gaining strength from the concrete representation of what I had to fight my illness for. Naturally, during Terrance's daily visits most of my questions were of Ivan. The picture of Ivan as a baby was more than just on my headboard, it was in my mind. I was thinking of Ivan as I was seeing him hourly. When I would go home, I couldn't wait to get him in my arms. Only parents who have experienced an involuntary separation from a dependent child can understand the overwhelming joy of being united again. Beyond keeping him clean, I was up to little more than cuddling, and that I did to no end.

Looking back, I see now that more damaging to Ivan than the sense of seeing him as a baby was my fear of pushing him to hard. Since I had often been classified by friends and business associates as aggressive, competitive, etc., I did not want to harm him by being pushy. Unfortunately, partly because of my illness, I went to the other extreme and did not push him at all. I understand now that infants and children need not pushing so much as prodding — that just as good mental health for adults comes with goal-directed behavior, young children must also be led, through our interactions with them, to reach the necessary goals of verbal, abstract reasoning, and motor skills that equip them to deal with their environment. They won't get there on their own. . . .

One thing bothered me. In order to help Ivan get where he should be in his relationship with other children, I had placed him in nursery school. I think his reactions or lack of them should have been brought immediately to my attention by his teacher. His deviant behavior could have solidified in the weeks before the official parent-teacher conferences.

The most tangible reason his excellent teacher did not approach me immediately, I believe, was her natural reluctance to categorize a child at the age of 3, when there are so many variables at play. Add to that the stigma attached to mental or emotional illnesses and their attendant computerized records that in our society limit career opportunities, some insurance coverage, and in general serve to discredit an individual throughout his or her entire life — and one can easily understand a hesitancy to suggest even the early signs of anything that serious.

When I frankly enumerated the probable reasons for his behavior, she agreed that his main problem was most likely the natural one of a child suffering along with his mother the adversities of a debilitating illness. She believed that her working with him and his participation in a classroom situation would be beneficial. Being in that school did help him tremendously.

I have learned. Now the first day of school I ask his teacher to inform me of any problem that is apparent in the first weeks or even days, no matter how minor. Minor problems, as I almost saw, can all too easily become major ones.

Because I had been on heavy medication and waking up was difficult, Ivan had often been the last child to enter his nursery school class — on time, but the last one. Knowing that most of us detest making an "entrance" into a social or other gathering, I made an extra effort to get him to school earlier to eliminate that shy, lonely arrival into a roomful of busily occupied preschoolers. Whenever Terrance was available, he would share the dropping off and picking up.

To continue to encourage Ivan to direct himself outward, I began to take him to the YWCA once a week for a play session while I took a course in physical fitness. This play time supplemented his 3 mornings a week at nursery school.

I joined a church, equally for my benefit, and placed Ivan in the Sunday School. Since I became the substitute teacher for his class, I was amused when I occasionally had to take over the class: He refused to let other children sit on my lap and was indignant if I read to them. I saw his possessiveness as a sign of health, and both we and the class managed to survive.

Terrance began to take Ivan fishing again — spring had come with its green lusciousness and appeal for rebirth and renewal. Whenever Ivan would catch a tiny fish, flipping with life, he and his father would carefully dismantle it from the hook and return it to the pond. The purpose of their "fishing" was not to catch food for suppers, but to introduce Ivan to the wonder of all things living.

Because I'm naturally more conscious of health now, I often think of how conditions are passed on from parents to children. Heart disease runs in families as do diabetes and an assortment of other medical ailments. There are many studies that prove the operation of genetic factors in the transmission of schizophrenia, but the importance of environment is also confirmed. Environment is important in any medical illness.

I'm reminded of the story of the alcoholic who had identical twin sons. One twin became a severe alcoholic. The other twin became a teetotaler —would never touch a drop. Fifty years later they were each asked why they had turned out the way they had. Their answers were identical: "With an alcoholic for a father, what else could I be?"

With all our physical and mental imperfections that we have inherited and will pass on to our children, one of the most important things we can do for them and for ourselves is to take stock of our medical backgrounds and program our lives accordingly. We must take care not to pass on to our children habits and a life style that exacerbate whatever genetic abnormality they might have. The deciding factor can be the way one lives, not necessarily embracing either of two extremes as the identical twins did, but to take a stand for your mental and physical health, realizing that both will function as you do.

In reflecting on all I have written, I wondered if I had over-reacted in my observations of Ivan — become too melodramatic, too guilt-ridden.

To clarify some of my beliefs, I asked my psychiatrist . . . if being with an ill schizophrenic mother could cause a child to pick up her symptoms. Her answer was, "Definitely, yes. Even though the child is not schizophrenic, he or she would still demonstrate a learned behavior pattern." [The cross-fostering study reviewed in Chapter 7 contradicts an uncritical acceptance of this conclusion.]

[My psychiatrist] also states that regarding schizophrenia and children as opposed to adults, if schizophrenia is genotypically present in children, it is more likely to appear phenotypically. Working with many schizophrenic children, she advises that the best way to prevent schizophrenia from appearing in a child predisposed to it is for parents to give that child — as with any of their offspring — wholesome, sincere, consistent love. [However, as demonstrated by the adoption strategies in Chapter 7, love is not always a defense that works. The myth of schizophrenogenic mothers or fathers, in any guise, should now be laid to rest by the research reviewed in this book.]

[My psychiatrist] affirms that a trait or personality behavior pattern is not in itself indicative of schizophrenia. She only makes such a diagnosis if the child or adult is rendered dysfunctional.

To allay any lingering fears that Ivan might have suffered permanent harm from my illness and many hospitalizations, or have some early signs of schizophrenia of which I was unaware, [my psychiatrist] gave him a preliminary evaluation to judge whether he warranted a more formal evaluation. She found him to be very bright, very creative, and enjoying a vigorous fantasy life appropriate for his age. She saw no evidence of any schizophrenia then (or even, in retrospect, during the most severe stages of my illness), and saw nothing to warrant further evaluation. Ivan enjoyed his session with her so much that now my only problem with my own checkup visits is making him understand that he can't have a session of his own.

I have learned much since that gloomy night 5 years ago when I sat at my kitchen table observing what I and my illness had done to Ivan. One thing I've learned is remarkable in its simplicity: The more secure a child is in his home life and in his parents' love, the more eager he is to leave that security and find it through his own achievements and in his relationship with himself and other people.

That very same night 5 years ago, I knew that Ivan was on his way to becoming the normal, healthy, and happy child he is today. When it was time for his favorite program, he didn't want to sit on my lap. He wanted his own chair. During a lull in the program he looked over at me, smiled — and waved.

References and Bibliography[1]

Chapter 1

Alexander, F. G. and Selesnick, S. T. (1966). *The History of Psychiatry*. New York: Harper & Row.

Allderidge, P. (1985). Bedlam: Fact or fiction? In W. F. Bynum, R. Porter and M. Shepherd (Eds.), *The Anatomy of Madness*, Vol. II (pp. 17–33). London: Tavistock.

Balzac, H. (1832/1966). Louis Lambert. In A. A. Stone and S. S. Stone (Eds.), *The Abnormal Personality Through Literature.* (pp. 52–64). Englewood Cliffs: Prentice-Hall.

*Bark, N. M. (1985). Did Shakespeare know schizophrenia? The case of poor mad Tom in *King Lear. British Journal of Psychiatry, 146*, 436–438.

*Berrios, G. E. (1987). Historical aspects of psychosis: 19th century issues. *British Medical Bulletin, 43*, 484–498.

Bleuler, E. (1908). Die Prognose der Dementia praecox (Schizophreniegruppe). *Allgemeine Zeitschrift fur Psychiatrie, 65*, 436–464.

Bleuler, E. *Dementia Praecox or the Group of Schizophrenias*. Leipzig: Deuticke, 1911. Republished: Trans. Joseph Zinkin. New York: International Universities Press, 1950.

---

[1]Items with an asterisk comprise background bibliography that was consulted but was not specifically referenced for this book.

Bynum, W. F. (1983). Psychiatry in its historical context. In M. Shepherd and O. L. Zangwill (Eds.), *Handbook of Psychiatry 1: General Psychopathology* (pp. 11–38). New York: Cambridge University Press.

Dix, D. (cf. Rothman, D. [1971]. *The Discovery of the Asylum.* Boston: Little, Brown & Co.).

Dobzhansky, T. (1968). On some fundamental concepts of Darwinian biology. In T. Dobzhansky, M. K. Hecht, and W. M. C. Steere (Eds.), *Evolutionary Biology*, Vol. 2 (pp. 1–34). New York: Appleton.

*Freud, S. (1896/1962). Further remarks on the neuro-psychoses of defence. *The Standard Edition of the Complete Psychological Works of Sigmund Freud*, Vol. 3. London: Hogarth, pp. 174–185.

*Grob, G. N. (1983). *Mental Illness and American Society 1875–1940.* Princeton: Princeton University Press.

Hare, E. H. (1983a). Was insanity on the increase? *British Journal of Psychiatry, 142*, 439–455.

Hare, E. H. (1983b). Epidemiological evidence for a viral factor in the aetiology of functional psychosis. In P. V. Morozov (Ed.), *Research on the Viral Hypothesis of Mental Disorders* (pp. 52–75). Basel: Karger.

*Hare, E. H. (1988). Schizophrenia as a recent disease. *British Journal of Psychiatry, 153*, 521–531.

Harris, H. (1963). *Garrod's Inborn Errors of Metabolism.* London: Oxford University Press (originally published in 1909).

Haslam, J. (1809/1976). *Observations on Madness and Melancholy.* New York: Arno Press.

*Hunter, R. A. and Macalpine, I. (1963). *Three Hundred Years of Psychiatry, 1535–1860.* London: Oxford University Press.

*Jeste, D. V., delCarmen, R., Lohr, J. B., and Wyatt, R. J. (1985). Did schizophrenia exist before the 18th century? *Comprehensive Psychiatry, 26*, 493–503.

*Karlsen, C. F. (1987). *The Devil in the Shape of a Woman— Witchcraft in Colonial New England.* New York: Norton.

Kraepelin, E. (1913/1919/1971). *Dementia Praecox and Paraphrenia.* (trans. R. M. Barclay, 1919). Huntington, New York: Robert E. Krieger Publishing Co.

*Kraepelin, E. (1987). *Emil Kraepelin Memoirs.* (trans. C. Wooding-Deene). Berlin: Springer-Verlag.

*Maher, W. B. and Maher, B. A. (1982). The ship of fools: Stultifera navis or ignis fatuus? *American Psychologist, 37*, 756–761.

Menninger, K. (1963). *The Vital Balance.* New York: Viking Press.

Morel, B. A. (1852). *Traite des Maladies Mentales.* Paris: Masson.

Pinel, P. (1801/1962). *A Treatise of Insanity.* New York: Hafner.

Rüdin, E. (1916). *Zur Vererbung und Neuentstehung der Dementia Praecox.* Berlin and New York: Springer-Verlag.

*Scheper-Hughes, N. (1979). *Saints, Scholars, and Schizophrenics.* Berkeley: University of California Press.

Schweitzer, A. (1913). *The Quest for the Historical Jesus* (2nd edition). (trans. W. Montgomery). London: A and C Black. Reprinted in 1948 as *The Psychiatric Study of Jesus* (trans. C. R. Joy). Boston: Beacon Press.

*Susser, M. (1973). *Causal Thinking in the Health Sciences.* London: Oxford University Press.

*Williams, P., Wilkinson, G., and Rawnsley, K. (Eds.) (1989). *The Scope of Epidemiological Psychiatry.* London: Routledge.

*Wilmer, H. A. and Scammon, R. E. (1954). Neuropsychiatric patients reported cured at St. Bartholomew's Hospital in the twelfth century. *Journal of Nervous and Mental Disease, 119,* 1–22.

Chapter 2

American Psychiatric Association (1980). *Diagnostic and Statistical Manual of Mental Disorders (DSM-III)* (3rd edition). Washington, D.C.: American Psychiatric Association.

American Psychiatric Association (1987). *Diagnostic and Statistical Manual of Mental Disorders (DSM-III-R)* (3rd edition, revised). Washington, D.C.: American Psychiatric Association.

Bleuler, E. (1911/1950). *Dementia Praecox or the Group of Schizophrenias.* (trans. Joseph Zinkin). New York: International Universities Press.

Bleuler, M. (1978). *The Schizophrenic Disorders: Long-Term Patient and Family Studies.* (trans. Siegfried M. Clemens). New Haven, Yale University Press.

*Buhrich, N., Cooper, D. A. and Freed, E. (1988). HIV infection associated with symptoms indistinguishable from functional psychosis. *British Journal of Psychiatry, 152,* 649–653.

*Chapman, L. J. and Chapman, J.P. (1987). The search for symptoms predictive of schizophrenia. *Schizophrenia Bulletin, 13,* 497–503.

*Claridge, G. (1988). Schizotypy and schizophrenia. In P. Bebbington and P. McGuffin (Eds.), *Schizophrenia: The Major Issues* (pp. 187–200). Oxford: Heinemann Medical Books.

Cooper, J. E., Kendell, R. E., Gurland, B. J., Sharpe, L., Copeland, J. R. M. and Simon, R. (1972). *Psychiatric Diagnosis in New York and London.* London: Oxford University Press.

*Cutting, J. (1985). *The Psychology of Schizophrenia.* Edinburgh: Churchill Livingstone.

*Cutting, J. (1987). The phenomenology of acute organic psychosis: Comparison with acute schizophrenia. *British Journal of Psychiatry, 151,* 324–332.

*Cutting, J. and Shepherd, M. (1987). *The Clinical Roots of the Schizophrenia Concept.* Cambridge: Cambridge University Press.

*Davison, K. and Bagley, C. R. (1969). Schizophrenic-like psychoses associated with organic disorders of the central nervous system: A review of the literature. In R. N. Herrington (Ed.), *Current Problems in Neuropsychiatry* (pp. 113–184). Ashford, Kent: Headley Bros.

Farmer, A. E., McGuffin, P. and Gottesman, I. I. (1987). Twin concordance for DSM-III schizophrenia: Scrutinizing the validity of the definition. *Archives of General Psychiatry, 44,* 634–641.

Fitzgerald, F. S. (1948). *The Crack-up.* New York: J. Laughlin.

*Folstein, S. and Rutter, M. (1977). Infantile autism: A genetic study of 21 twin pairs. *Journal of Child Psychology and Psychiatry, 18,* 297–321.

Gottesman, I. I. and Shields, J. (1972). *Schizophrenia and Genetics: A Twin Study Vantage Point.* New York: Academic Press.

Greenberg, J. (1964). *I Never Promised You a Rose Garden.* New York: Holt, Rinehart and Winston.

*Hanson, D. R. and Gottesman, I. I. (1976). The genetics, if any, of infantile autism and childhood schizophrenia. *Journal of Autism and Childhood Schizophrenia, 6,* 209–234.

Jaspers, K. (1913/1963). *General Psychopathology.* Trans J. Hoenig and M. W. Hamilton from the seventh German edition. Chicago: Chicago University Press.

Kanner, L. (1943). Autistic disturbances of affective contact. *Nervous Child, 2,* 217–250.

Kuriansky, J. B., Deming, W. E. and Gurland, B. J. (1974). On trends in the diagnosis of schizophrenia. *American Journal of Psychiatry, 131,* 402–408.

*Lenzenweger, M. F. and Loranger, A. W. (1989). Psychosis proneness and clinical psychopathology: Examination of the correlates of schizotypy. *Journal of Abnormal Psychology, 93,* 3–8.

McGuffin, P., Farmer, A. E. and Gottesman, I. I. (1987). Is there really a split in schizophrenia? the genetic evidence. *British Journal of Psychiatry, 150,* 581–592.

*McKusick, V. A. (1969). On lumpers and splitters, or the nosology of genetic disease. *Perspectives on Biology and Medicine, 12,* 298–310.

*Meehl, P. E. (1986). Diagnostic taxa as open concepts: Metatheoretical and statistical questions about reliability and construct validity in the grand strategy of nosological revision. In T. Millon and G. Klerman (Eds.), *Contemporary Directions in Psychopathology: Toward the DSM-IV* (pp. 215–231). New York: Guilford Press.

*North, C. and Cadoret, R. (1981). Diagnostic discrepancy in personal accounts of patients with "schizophrenia." *Archives of General Psychiatry, 38,* 133–137.

*Oltmanns, T. F. and Maher, B. A. (1988). *Delusional Beliefs.* New York: John Wiley & Sons.

*Petty, L. K., Ornitz, E. M., Michelman, J. D. and Zimmerman, E. G. (1984). Autistic children who become schizophrenic. *Archives of General Psychiatry, 41,* 129–135.

*Propping, P. (1983). Genetic disorders presenting as "schizophrenia." *Human Genetics, 65,* 1–19.

Romano, J. (1977). On the nature of schizophrenia: Changes in the observer as well as the observed (1932–77). *Schizophrenia Bulletin, 3,* 532–559.

*Rothblum, E. D., Solomon, L. J. and Albee, G. W. (1986). A sociopolitical perspective of DSM-III. In T. Millon and G. Klerman (Eds.), *Contemporary Directions in Psychopathology: Toward the DSM-IV* (pp. 167–189). New York: Guilford Press.

*Rumsey, J. M., Rappaport, J. L. and Sceery, W. R. (1985). Autistic children as adults: Psychiatric, social, and behavioral outcomes. *Journal of the American Academy of Child Psychiatry, 24,* 465–473.

Schneider, K. (1959). *Clinical Psychopathology* (trans. M. W. Hamilton). New York: Grune and Stratton.

Sheehan, S. (1982). *Is There No Place on Earth for Me?* New York: Random House.

Sims, A. (1988). *Symptoms of the Mind: An Introduction to Descriptive Psychopathology.* London: Bailiere Tindall.

*Smalley, S. L., Asarnow, R. F. and Spence, M. A. (1988). Autism and genetics. *Archives of General Psychiatry, 45,* 953–961.

Thigpen, C. H. and Cleckley, H. M. (1957). *The Three Faces of Eve.* Kingsport, Conn.: Kingsport Press.

Vonnegut, M. (1975). *The Eden Express.* New York: Praeger.

Wing, J. K., Cooper, J. E. and Sartorius, N. (1974). *Measurement and Classification of Psychiatric Symptoms: An Instructional Manual for the PSE and Catego Program.* London: Cambridge University Press.

Wing, L. (Ed.) (1988). *Aspects of Autism: Biological Research.* London: Gaskell.

*Wolfe, S. and Cull, A. (1986). "Schizoid" personality and antisocial conduct: A retrospective case note study. *Psychological Medicine, 16,* 677–687.

World Health Organization (1973). *The International Pilot Study of Schizophrenia,* Vol. 1. Geneva: World Health Organization.

Chapter 3

*Macalpine, I. and Hunter, R. (Eds.) (1955). *Daniel Paul Schreber: Memoirs of My Nervous Illness.* London: William Dawson (originally published in 1903).

*North, C. (1987). *Welcome Silence: My Triumph Over Schizophrenia.* New York: Simon and Schuster.

*Porter, R. (1987). *A Social History of Madness — The World Through the Eyes of the Insane.* New York: Weidenfeld & Nicolson.

*Sechehaye, M. (1951). *Autobiography of a Schizophrenic Girl.* New York: Grune and Stratton.

*West, D. J. and Walk, A. (Eds.). (1977). *Daniel McNaughton — His Trial and the Aftermath.* Ashford, Kent: Gaskell Books.

Chapter 4

*Barrett, J. and Rose, R. M. (Eds.) (1986). *Mental Disorders in the Community.* New York: Guilford.

Cox, P. R. (1970). *Demography.* Cambridge: Cambridge University Press.

Dublin, L. I., Lotka, A. J. and Spiegelman, M. (1949). *Length of Life.* New York: Ronald Press.

Eaton, W. W. (1985). Epidemiology of schizophrenia. *Epidemiological Review, 7,* 105–126.

Essen-Möller, E., Larsson, H., Uddenberg, C. E. and White, G. (1956). Individual traits and morbidity in a Swedish rural population. *Acta Psychiatrica et Neurologica Scandinavica, Supplement 100.*

Faris, R. E. L. and Dunham, H. W. (1939). *Mental Disorders in Urban Areas.* Chicago: University of Chicago Press.

*Flekkøy, K. (1987). Epidemiologie und Genetik. In K. P. Kisker, H. Lauter, J.-E. Meyer, C. Müller and E. Stömgren (Eds.), *Psychiatrie der Gegenwart 4* (pp. 119–153). Berlin: Springer Verlag.

Fremming, K. (1951). *Expectation of Mental Infirmity: Occasional Papers on Eugenics* (No. 7). London: Cassell.

Frost, W. H. (1936). *Snow on Cholera* (reprint of John Snow, 1854). London: Oxford University Press.

Goldberg, E. M. and Morrison, S. L. (1963). Schizophrenia and social class. *British Journal of Psychiatry, 109*, 785–802.

Goldberger, J. (1914). The etiology of pellagra: The significance of certain epidemiological observations with respect thereto. *Public Health Reports, 29*, 1683–1686.

*Hare, E. H. (1959). The origin and spread of dementia paralytica. *Journal of Mental Science, 105*, 594–626.

Hare, E. H. (1988). Schizophrenia as a recent disease. *British Journal of Psychiatry, 153*, 521–531.

*Hollingsworth, T. H. (1969). *Historical Demography*. Ithaca: Cornell University Press.

Jablensky, A. (1988). Epidemiology of schizophrenia. In P. Bebbington and P. McGuffin (Eds.), *Schizophrenia: The Major Issues* (pp. 19–35). London: Heinemann Medical Books.

*Janssens, P. A. (1970). *Paleopathology: Diseases and Injuries of Prehistoric Man*. London: John Baker.

*Keyfitz, N. and Flieger, W. (1971). *Population — Facts and Methods of Demography*. San Francisco: W. H. Freeman.

Loranger, A. (1984). Sex differences in age at onset of schizophrenia. *Archives of General Psychiatry, 41*, 157–161.

McNeill, W. H. (1976). *Plagues and Peoples*. New York: Anchor Books.

*Munk-Jorgensen, P. (1986). Decreasing first-admission rates of schizophrenia among males in Denmark from 1970–1984. Changing diagnostic patterns? *Acta Psychiatrica Scandinavica, 74*, 645–650.

Ødegaard, Ø. (1952). La Genetique dans la Psychiatrie. *Proceedings First World Congress of Psychiatry*, Paris, 1950. Paris: Hermann. *Comptes rendus*, VI, pp. 84–90.

Ødegaard, Ø. (1971). Hospitalized psychoses in Norway: Time trends 1926–1965. *International Journal of Social Psychiatry, 6*, 53–58.

*Ødegaard, Ø. (1972). Epidemiology of the psychoses. In K. P. Kisker, J. E. Meyer, C. Müller and E. Strömgren (Eds.), *Psychiatrie der Gegenwart*, Vol. 2 (pp. 213–258). Berlin: Springer-Verlag.

*Ødegaard, Ø. (1975). Social and ecological factors in the etiology, outcome, treatment and prevention of mental disorders. In K. P. Kisker, J. E. Meyer, C. Müller and E. Strömgren (Eds.), *Psychiatrie der Gegenwart*, Vol. 3. (pp. 151–198). Berlin: Springer-Verlag.

*Roe, D. A. (1973). *A Plague of Corn: The Social History of Pellagra*. Ithaca: Cornell University Press.

Rosenstein, M. J., Milazzo-Sayre, L. J., and Mandershied, R. W. (1989). Care of persons with schizophrenia: A statistical profile. *Schizophrenia Bulletin, 15*, 45–58.

Sartorious, N., Jablensky, A., Korten, A., Ernberg, G., Anker, M., Cooper, J. E. and Day, R. (1986). Early manifestations and first-contact incidence of schizophrenia in different cultures. *Psychological Medicine, 16*, 909–928.

Saugstad, L. F. (1989). Social class, marriage, and fertility in schizophrenia. *Schizophrenia Bulletin, 15*, 9–43.

Slater, E. and Cowie, V. (1971). *The Genetics of Mental Disorders*. London: Oxford University Press.

Spies, T. D., Cooper, C. and Blankenhorn, M. A. (1936). The use of nicotinic acid in the treatment of pellagra. *Journal of the American Medical Association, 110*, 622–627.

*Steadman, H. J., Monahan, J., Duffee, B., Hartstone, E. and Robbins, P. C. (1984). The impact of state mental hospital deinstitutionalization on United States prison populations, 1968–1978. *Journal of Criminal Law and Criminology, 75*, 474–490.

Torrey, E. F. (1980). *Schizophrenia and Civilization*. New York: Jason Aronson.

*Torrey, E. F. (1988). *Nowhere to Go—The Tragic Odyssey of the Homeless Mentally Ill*. New York: Harper & Row.

U. S. Bureau of The Census, *Statistical Abstract of the United States: 1988* (108th edition). Washington, D.C.

*Wrigley, E. A. and Schofield, R. S. (1982). *Population History of England 1541–1871*. Cambridge, Mass.: Harvard University Press.

Chapters 5 and 6

*Berg, K. (1987). Genetic risk factors for atherosclerotic disease. In F. Vogel and K. Sperling (Eds.), *Human Genetics* (pp. 326–335). Berlin: Springer-Verlag.

Bleuler, M. (1978). *The Schizophrenic Disorders: Long-Term Patient and Family Studies.* (trans. Siegfried M. Clemens). New Haven: Yale University Press.

Buchsbaum, M. S., Mirsky, A. F., DeLisi, L. E., Morihisa, J., Karson, C. N., Mendelson, W. B., King, A. C., Johnson, J. and Kessler, R. (1984). The Genain quadruplets: Electrophysiological, positron emission, and x-ray tomographic studies. *Psychiatry Research, 13*, 95–108.

Bulmer, M. G. (1970). *The Biology of Twinning in Man.* Oxford: Clarendon Press.

Caplan, A. L. (Ed.). (1978). *The Sociobiology Debate.* New York: Harper & Row.

Crow, T. J. (1985). The two-syndrome concept: Origins and current states. *Schizophrenia Bulletin, 11*, 471–486.

Crow, T. J. (1986). The continuum of psychosis and its implications for the structure of the gene. *British Journal of Psychiatry, 149*, 419–429.

*Crow, T. J. (1988). Aetiology of psychosis: The way ahead. In P. Bebbington and P. McGuffin (Eds.), *Schizophrenia: The Major Issues* (pp. 127–143). London: Heinemann Medical Books.

Eisenberg, L. (1968). The interaction of biological and experiential factors in schizophrenia. In D. Rosenthal and S. S. Kety (Eds.), *The Transmission of Schizophrenia* (pp. 403–409). Oxford: Pergamon.

Eisenberg, L. (1988). The relationship between psychiatric research and public-policy. *British Journal of Psychiatry, 153*, 21–29.

Essen-Möller, E. (1941). Psychiatrische Untersuchungen an einer Serie von Zwillingen. *Acta Psychiatrica et Neurologica Scandinavica, Supplement 23.*

Farmer, A. E., McGuffin, P. and Gottesman, I. I. (1987). Twin concordance for DSM-III schizophrenia: Scrutinizing the validity of the definition. *Archives of General Psychiatry, 44*, 634–641.

Feighner, J. P., Robins, E., Guze, S. B., Woodruff, R. A., Winokur, G. and Munoz, R. (1972). Diagnostic criteria for use in psychiatric research. *Archives of General Psychiatry, 26*, 57–63.

Fischer, M. (1971). Psychoses in the offspring of schizophrenic monozygotic twins and their normal co-twins. *British Journal of Psychiatry, 118*, 43–52.

Fischer, M. (1973). Genetic and environmental factors in schizophrenia. *Acta Psychiatrica Scandinavica, Supplement 238.*

Fromm-Reichmann, F. (1948). Notes on the development of treatments of schizophrenics by psychoanalytic psychotherapy. *Psychiatry, 2*, 263–273.

Galton, F. (1875). The history of twins, as a criterion of the relative powers of nature and nurture. *Fraser's Magazine, 12*, 566–576.

Gottesman, I. I. and Bertelsen, A. (1989). Confirming unexpressed genotypes for schizophrenia: Risks in the offspring of Fischer's Danish identical and fraternal discordant twins. *Archives of General Psychiatry, 46*, 867–872.

Gottesman, I. I. and Bertelsen, A. (1989). Dual mating studies in psychiatry — offspring of inpatients with examples from reactive (psychogenic) psychoses. *International Review of Psychiatry, 1*, 287–296.

Gottesman, I. I. and Carey, G. (1983). Extracting meaning and direction from twin data. *Psychiatric Developments, 1*, 398–404.

Gottesman, I. I., McGuffin, P. and Farmer, A. E. (1987). Clinical genetics as clues to the "real" genetics of schizophrenia (a decade of modest gains while playing for time). *Schizophrenia Bulletin, 13*, 23–47.

Gottesman, I. I. and Shields, J. (1972). *Schizophrenia and Genetics: A Twin Study Vantage Point.* New York: Academic Press.

Gottesman, I. I. and Shields, J. (1976a). A critical review of recent adoption, twin, and family studies of schizophrenia: Behavioral genetics perspective. *Schizophrenia Bulletin, 2*, 360–401.

Gottesman, I. I. and Shields, J. (1976b). Rejoiner: toward optimal arousal and away from origin din. *Schizophrenia Bulletin, 2*, 447–453.

Gottesman, I. I. and Shields, J. (with assistance of D. R. Hanson) (1982). *Schizophrenia: The Epigenetic Puzzle.* New York: Cambridge University Press.

Inouye, E. (1961). Similarity and dissimilarity of schizophrenia in twins. *Proceedings Third International Congress of Psychiatry, 1*, 524–530 (Montreal: University of Toronto Press, 1963).

Inouye, E. (1972). Monozygotic twins with schizophrenia reared apart in infancy. *Japanese Journal of Human Genetics, 16*, 182–190.

Jackson, D. D. (1960). A critique of the literature on the genetics of schizophrenia. In D. D. Jackson (Ed.), *The Etiology of Schizophrenia* (pp. 37–87). New York: Basic Books.

Kallmann, F. J. (1938). *The Genetics of Schizophrenia.* New York: Augustin.

Kallmann, F. J. (1946). The genetic theory of schizophrenia: An analysis of 691 schizophrenic twin index families. *American Journal of Psychiatry, 103*, 309–322.

Kay, D. W. K. and Lindelius, R. (1970). Morbidity risks for schizophrenia among parents, siblings, probands' children, and siblings' children. *Acta Psychiatrica Scandinavica, Supplement 216*, 86–88.

Kendler, K. S. (1988). Familial aggregation of schizophrenia and schizophrenia spectrum disorders. *Archives of General Psychiatry, 45*, 377–383.

Kendler, K. S. and Robinette, C. D. (1983). Schizophrenia in the National Academy of Sciences-National Research Council Twin Registry: A 16-year update. *American Journal of Pscyhiatry, 140*, 1551–1563.

Kringlen, E. (1967). *Heredity and Environment in the Functional Psychoses.* London: Heinemann Medical Books.

Luxenburger, H. (1928). Vorlaufiger Bericht Uber psychiatrische Serienuntersuchungen an Zwillingen. *Zeitschrift fur die gesamte Neurologie und Psychiatrie, 116*, 297–326.

McGuffin, P., Farmer, A. E., Gottesman, I. I., Murray, R. M. and Reveley, A. M. (1984). Twin concordance for operationally defined schizophrenia. *Archives of General Psychiatry, 41*, 541–545.

McGuffin, P. and Gottesman, I. I. (1985). Genetic influences on normal and abnormal development. In M. Rutter and L. Hersov (Eds.), *Child and Adolescent Psychiatry: Modern Approaches* (2nd edition) (pp. 17–33). Oxford: Blackwell Scientific Publications.

McGuffin, P., Reveley, A. M. and Holland, A. (1982). Identical triplets: Nonidentical psychosis? *British Journal of Psychiatry, 140*, 1–6.

McKusick, V. A. (1990). *Mendelian Inheritance in Man: Catalogs of Autosomal Dominant, Autosomal Recessive, and X-Linked Phenotypes* (9th edition). Baltimore: Johns Hopkins University Press.

Meehl, P. E. (1972). Specific genetic etiology, psychodynamics, and therapeutic nihilism. *International Journal of Mental Health, 1*, 10–27.

Menninger, K. (1962). *The Vital Balance*. New York: Viking Press.

Mosher, L. R., Pollin, W. and Stabenau, J. R. (1971). Identical twins discordant for schizophrenia: Some relationships between identification, thinking styles, psychopathology and dominance submissiveness. *British Journal of Psychiatry, 118*, 29–42.

*Pollin, W., Allen, M. G., Hoffer, A., Stabenau, J. R. and Hrubec, Z. (1969). Psychopathology in 15,909 pairs of veteran twins: Evidence for a genetic factor in the pathogenesis of schizophrenia and its relative absence in psychoneurosis. *American Journal of Psychiatry, 126*, 597–610.

Rosanoff, A. J., Handy, L. M., Plesset, I. R. and Brush, S. (1934). The etiology of so-called schizophrenic psychoses with special reference to their occurrence in twins. *American Journal of Psychiatry, 91*, 247–286.

Rosenthal, D. (1960). Confusion of identity and the frequency of schizophrenia in twins. *Archives of General Psychiatry, 3*, 297–304.

Rosenthal, D. (1961). Sex distribution and the severity of illness among samples of schizophrenic twins. *Journal of Psychiatric Research, 1*, 26–36.

Rosenthal, D. (1962). Problems of sampling and diagnosis in the major twin studies of schizophrenia. *Journal of Psychiatric Research, 1*, 116–134.

Rosenthal, D. (Ed.) et al. (1963). *The Genain Quadruplets*. New York: Basic Books.

Rosenthal, D. and Kety, S. S. (Eds.) (1968). *The Transmission of Schizophrenia*. Oxford: Pergamon.

Rotter, J. I., King, R. and Motulsky, A. (Eds.) (1990). *The Genetics of Common Disorders.* New York: Oxford University Press (in press).

Rüdin, E. (1916). *Zur Vererbung und Neuentstehung der Dementia Praecox.* Berlin and New York: Springer-Verlag.

Scharfetter, C. (1972). Studies of heredity in symbiontic psychoses. *International Journal of Mental Health, 1,* 116–123.

Schneider, K. (1959). *Clinical Psychopathology* (trans. M. W. Hamilton). New York: Grune and Stratton.

Schulz, B. (1932). Zur Erbpathologie der Schizophrenie. *Zeitschrift für die gesamte Neurologie und Psychiatrie, 143,* 175–293.

Shields, J. (1968). Summary of the genetic evidence. In D. Rosenthal and S. S. Kety (Eds.), *The Transmission of Schizophrenia* (pp. 95–126). Oxford: Pergamon.

Shields, J., Gottesman, I. I. and Slater, E. (1967). Kallmann's 1946 twin study in the light of new information. *Acta Psychiatrica Scandinavica, 43,* 385–396.

Slater, E. (with the assistance of Shields, J.) (1953). Psychotic and neurotic illnesses in twins. *Medical Research Council Special Report Series No. 278.* London: Her Majesty's Stationery Office.

Smith, C. (1974). Concordance in twins: Methods and interpretation. *American Journal of Human Genetics, 26,* 454–466.

Tienari, P. (1975). Schizophrenia in Finnish male twins. In M. H. Lader (Ed.), *Studies of Schizophrenia* (pp. 29–35). Ashford, Kent: Headley Brothers.

Vogel, F. and Motulsky, A. G. (1986). *Human Genetics* (2nd edition). Berlin: Springer-Verlag.

Chapter 7

Bateson, G., Jackson, D. D., Haley, J. and Weakland, J. (1956). Toward a theory of schizophrenia. *Behavioral Science, 1,* 251–264.

Bertalanffy, L. von. (1966). General systems theory and psychiatry. In S. Arieti (Ed.), *American Handbook of Psychiatry*, Vol 3. (pp. 705–721). New York: Basic Books.

Burgess, E. W. (1926). The family as a unit of interacting personalities. *Family, 7,* 3–9.

Dupont, A. (1983). A national psychiatric case register as a tool for mental health planning, research, and administration: The Danish model. In E. M. Laska, M. S. Gulbinat and D. A. Regier (Eds.), *Information Support to Mental Health Programs* (pp. 257–274). New York: Human Sciences Press.

*Grove, W. M. (1983). Comment on Lidz and associates' critique of the Danish-American studies of the offspring of schizophrenic parents. *American Journal of Psychiatry, 140*, 998–1002.

Heston, L. L. (1966). Psychiatric disorders in foster home reared children of schizophrenic mothers. *British Journal of Psychiatry, 112*, 819–825.

Jacobs, T. (Ed.) (1986). *Family Interaction and Psychopathology: Theories, Methods and Findings.* New York: Plenum.

Kallmann, F. J. (1938). *The Genetics of Schizophrenia.* New York: Augustin.

Kendler, K. S. and Gruenberg, A. M. (1984). An independent analysis of the Copenhagen sample of the Danish adoption study of schizophrenia: VI. The pattern of psychiatric illness, as defined by DSM-III in adoptees and relatives. *Archives of General Psychiatry, 41*, 555–564.

Kendler, K. S., Masterson, C. C. and Davis, K. L. (1985). Psychiatric illness in first-degree relatives of patients with paranoid psychosis, schizophrenia, and medical illness. *British Journal of Psychiatry, 147*, 524–531.

Kety, S. S. (1988). Schizophrenic illness in the families of schizophrenic adoptees: Findings from the Danish national sample. *Schizophrenia Bulletin, 14*, 217–222.

Kety, S. S., Rosenthal, D., Wender, P. H., Schulsinger, F. and Jacobsen, B. (1978). The biological and adoptive families of adopted individuals who become schizophrenic. In L. C. Wynne, R. L. Cromwell and S. Matthysse (Eds.), *The Nature of Schizophrenia* (pp. 25–37). New York: John Wiley & Sons.

King, D. J. and Cooper, S. J. (1989). Viruses, immunity, and mental disorder. *British Journal of Psychiatry, 154*, 1–7.

*Litz, T., Blatt, S. and Cook, B. (1981). Critique of the Danish-American studies of the adopted-away offspring of schizophrenic parents. *American Journal of Psychiatry, 138*, 1063–1068.

Lowing, P. A., Mirsky, A. F. and Pereira, R. (1983). The inheritance of schizophrenic spectrum disorders: A reanalysis of the Danish adoptee study data. *American Journal of Psychiatry, 140*, 1167–1171.

Rosenthal, D. (1975). Discussion: The concept of schizophrenic disorders. In R. R. Fieve, D. Rosenthal and H. Brill (Eds.), *Genetic Research in Psychiatry* (pp. 199–208). Baltimore: Johns Hopkins University Press.

Rosenthal, D., Wender, P. H., Kety, S. S., Schulsinger, F., Welner, J. and Ostergaard, L. (1968). Schizophrenic's offspring reared in adoptive homes. In D. Rosenthal and S. S. Kety (Eds.), *The Transmission of Schizophrenia* (pp. 377–391). Oxford: Pergamon.

*Schanda, H., Berner, P., Gabriel, E., Kornberger, M.-L. and Kufferle, B. (1983). The genetics of delusional psychoses. *Schizophrenia Bulletin, 9*, 563–570.

Singer, M. T. and Wynne, L. C. (1965). Thought disorder and family relations of schizophrenics: IV. Results and implications. *Archives of General Psychiatry, 12*, 201–212.

Tienari, P., Sorri, A., Lahti, I., Naarla, M., Wahlberg, K. E., Moring, J., Pahjola, J. and Wynne, L. C. (1987). Interaction of genetic and psychosocial factors in schizophrenia. *Schizophrenia Bulletin, 13*, 477–484.

Wender, P. H., Rosenthal, D., Kety, S. S., Schulsinger, F. and Welner, J. (1974). Cross-fostering: A research strategy for clarifying the role of genetic and experiential factors in the etiology of schizophrenia. *Archives of General Psychiatry, 30*, 121–128.

Chapter 8

Bebbington, P. and Kuipers, L. (1988). Social influences on schizophrenia. In P. Bebbington and P. McGuffin (Eds.), *Schizophrenia—The Major Issues* (pp. 201–225). London: Heinemann Medical Books.

Brown, G. W. (1985). The discovery of expressed emotion: Induction or deduction? In J. Leff and C. Vaughn (Eds.), *Expressed Emotion in Famlies: Its Significance for Mental Illness* (pp. 7–25). New York: Guilford Press.

Brown, G. W., Birley, J. L. T. and Wing, J. K. (1972). Influence of family life on the course of schizophrenic disorders: A replication. *British Journal of Psychiatry, 121*, 241–258.

Centers for Disease Control (1988). Serum, 2,3,7,8-tetrachlorodibenzo-p-dioxin levels in U.S. Army Vietnam-era veterans. *Journal of the American Medical Association, 260*, 1249–1254.

Cooper, B. (1987). Mental disorder as reaction: The history of a psychiatric concept. In H. Katschnig (Ed.), *Life Events and Psychiatric Disorders* (pp. 1–32). Cambridge: Cambridge University Press.

Day, R. (1986). Social stress and schizophrenia: From the concept of recent life events to the notion of toxic environments. In G. D. Burrows and T. R. Norman (Eds.), *Handbook of Studies on Schizophrenia* (pp. 71–82). Amsterdam: Elsevier.

Dohrenwend, B. P., Shrout, P. E., Link, B. G. and Skodol, A. E. (1987). Social and psychological risk factors for episodes of schizophrenia. In H. Häfner, W. F. Gattaz and W. Janzarik (Eds.), *Search for the Causes of Schizophrenia* (pp. 275–296). Berlin: Springer-Verlag.

Eitinger, L. (1964). *Concentration camp survivors in Norway and Israel.* Oslo: Universitetsforlaget.

Gottesman, I. I. and Shields, J. (with the assistance of D. R Hanson) (1982). *Schizophrenia: The Epigenetic Puzzle.* New York: Cambridge University Press.

Hogarty, G. E., McEvoy, J. P., Munetz, M., DiBarry, A. L., Bartone, P., Cather, R., Cooley, S. J., Ulrich, R. F., Carter, M. C. and Madonia, M. J. (1988). Dose of fluphenazine, familial expressed emotion, and outcome in schizophrenia. *Archives of General Psychiatry, 45,* 797–805.

Leff, J. and Vaughn, C. (1985). *Expressed Emotion in Families: Its Significance for Mental Illness.* New York: Guilford Press.

Leff, J., Berkowitz, R., Sharit, N., Strachan, A., Glass, I. and Vaughn, C. (1989). A trial of family therapy versus a relative's group for schizophrenia. *British Journal of Psychiatry, 154,* 58–66.

McNeil, T. F. (1987). Perinatal influences in the development of schizophrenia. In H. Helmchen and F. A. Henn (Eds.), *Biological Perspectives of Schizophrenia* (pp. 125–138). New York: John Wiley & Sons.

Rosenthal, D. and Kety, S. S. (Eds.) (1968). *The Transmission of Schizophrenia.* Oxford: Pergamon.

Schofield, W. and Balian, L. A. (1959). A comparative study of the personal histories of schizophrenic and nonpsychiatric patients. *Journal of Abnormal and Social Psychology, 59,* 216–225.

Singer, M. T. and Wynne, L. C. (1965). Thought disorder and family relations of schizophrenics: IV. Results and implications. *Archives of General Psychiatry, 12,* 201–212.

Slater, E. (1943). The neurotic constitution: A statistical study of two thousand neurotic soldiers. *Journal of Neurology and Psychiatry, 6,* 1–16. Reprinted in J. Shields and I. I. Gottesman (Eds.) (1971). *Man, Mind and Heredity* (pp. 191–215). Baltimore: Johns Hopkins University Press.

Steinberg, H. and Durell, J. (1968). A stressful situation as a precipitant of schizophrenic symptoms. *British Journal of Psychiatry, 114,* 1097–1105.

Strachan, A. M. (1986). Family intervention for the rehabilitation of schizophrenia: Toward protection and coping. *Schizophrenia Bulletin, 12,* 678–698.

*Svendsen, B. B. (1952). *Psychiatric Morbidity Among Civilians in Wartime.* Copenhagen: Munksgaard.

Tarrier, N., Barraclough, C., Proceddu, K. and Watts, S. (1988). The assessment of psychological reactivity to the expressed emotion of the relatives of schizophrenic patients. *British Journal of Psychiatry, 152,* 618–624.

Tienari, P. (1963). Psychiatric illness in identical twins. *Acta Psychiatrica Scandinavica, Supplement 171.*

Tienari, P. (1966). On intrapair differences in male twins with special reference to dominance-submissiveness. *Acta Psychiatrica Scandinavia, Supplement 188.*

Vaughn, C. and Leff, J. P. (1976). The influence of family and social factors on the course of psychiatric illness. *British Journal of Psychiatry, 129,* 125–137.

Wagner, P. S. (1946). Psychiatric activities during the Normany offensive June 20–August 20, 1944. *Psychiatry, 9,* 341–364.

Wynne, L. C., Cromwell, R. L. and Matthysse, S. (Eds.) (1978). *The Nature of Schizophrenia: New Approaches to Research and Treatment.* New York: John Wiley & Sons.

Chapter 9

*Anon. (1987). *Medication for Mental Illness.* Rockville, NIMH. (DHHS Publication No. [ADM] 87-1509).

*Estroff, S. E. (1981). *Making It Crazy: An Ethnography of Psychiatric Clients in an American Community.* Berkeley: University of California Press.

*Falloon, A. R. H., Boyd, J. L. and McGill, C. W. (1984). *Family Care of Schizophrenia.* New York: Guilford.

*Gibbons, J. S., Horn, S. H., Powell, J. M. and Gibbons, J. L. (1984). Schizophrenic patients and their families. *British Journal of Psychiatry, 144,* 70–77.

*Gunderson, J. G., Frank, A. F., Katz, H. M., Vannichelli, M. L., Frosch, J. P. and Knapp, P. H. (1984). Effects of psychotherapy in schizophrenia: II. Comparative outcome of two forms of treatment. *Schizophrenia Bulletin, 10,* 564–598.

*Hatfield, A. B. (1981). The family as partner in the treatment of mental illness. *Hospital and Community Psychiatry, 20,* 338–340.

*Jones, K. and Poletti, A. (1985). Understanding the Italian experience. *British Journal of Psychiatry, 146,* 341–347.

*Kane, J. M. (1987). Treatment of schizophrenia. *Schizophrenia Bulletin, 13,* 133–156.

*Klerman, G. L. (1984). Ideology and science in the individual psychotherapy of schizophrenia. *Schizophrenia Bulletin, 10,* 608–612.

*Kreisman, D. E. and Joy, V. D. (1974). Family response to the mental illness of a relative: A review of the literature. *Schizophrenia Bulletin, 10,* 34–57.

*Lipton, M. A., Mailman, R. B. and Nemeroff, C. B. (1979). Vitamins, megavitamin therapy, and the nervous system. In R. J. Wurtman and J. J. Wurtman (Eds.), *Nutrition and the Brain*, Vol. 3, (pp. 183–264). New York: Raven.

*Monahan, J. (1988). Risk assessment of violence among the mentally disordered: Generating useful knowledge. *International Journal of Law and Psychiatry, 11*, 249–257.

*Sheehan, S. (1982). *Is There No Place on Earth for Me?* New York: Random House.

*Tornatore, F. L., Sramek, J. J., Okeya, B. L. and Pi, E. H. (1987). *Reactions to Psychotropic Medications*. New York: Plenum.

*Torrey, E. F. (1988). *Surviving Schizophrenia: A Family Manual* (Revised). New York: Harper & Row.

*Vonnegut, M. (1974). Why I want to bite R. D. Laing. *Harpers, 248*, 90–92.

*Walsh, M. (1985). *Schizophrenia: Straight Talk for Families and Friends*. New York: William Morrow.

*Wasow, M. (1982). *Coping with Schizophrenia: A Survival Manual for Parents, Relatives and Friends*. Palo Alto: Science and Behavior Books.

*Wasow, M. (1986). The need for asylum for the chronically mentally ill. *Schizophrenia Bulletin, 12*, 162–167.

Chapter 10

Allebeck, P. (1989). Schizophrenia: A life-shortening disease. *Schizophrenia Bulletin, 15*, 81–89.

Andreasson, S., Allebeck, P., Engstrom, A. and Rydberg, U. (1987). Cannabis and schizophrenia: A longitudinal study of Swedish conscripts. *Lancet, 2*, 1483–1485.

*Applebaum, P. S. (1987). Legal aspects of violence by psychiatric patients. *Annual Review of Psychiatry, 6*, 549–564.

Arendt, H. (1973). *The Origins of Totalitarianism*. New York: Harcourt Brace Jovanovich.

Baldwin, J. A. (1979). Schizophrenia and physical disease. *Psychological Medicine, 9*, 611–619.

*Berkowitz, R. I., Coustan, D. R. and Mochizuki, T. K. (Eds.). (1981). *Handbook for Prescribing Medication During Pregnancy*. Boston: Little, Brown & Co.

Bleuler, M. (1978). *The Schizophrenic Disorders: Long-Term Patient and Family Studies*. (trans. Siegfried M. Clemens). New Haven: Yale University Press.

*Chapman, L. J. (1990). Meehl's theory of schizotaxia, schizotypy, and schizophrenia. *Journal of Personality Disorders, 4:* 111–115.

Dickens, B. M. (1987). Comparative law and legislation on eugenic sterilization and selective abortion. In F. Vogel and K. Sperling (Eds.), *Human Genetics* (pp. 673–682). Berlin: Springer-Verlag.

Dupont, A., Moeller-Jensen, O., Strömgren, E. and Jablensky, A. (1986). Incidence of cancer in patients diagnosed as schizophrenic in Denmark. In G. H. M. M. ten Horn, R. Giel, W. Gulbinat and J. H. Henderson (Eds.), *Psychiatric Case Registers in Public Health* (pp. 229–239). Amsterdam: Elsevier.

Erlenmeyer-Kimling, L. (1978). Fertility of psychotics: Demography. In R. Cancro (Ed.), *Annual Review of the Schizophrenic Syndrome*, Vol. 5 (pp. 298–333). New York: Brunner/Mazel.

Gibbons, J. S., Horn, S. H., Powell, J. M. and Gibbons, J. L. (1984). Schizophrenic patients and their families. *British Journal of Psychiatry, 144,* 70–77.

Gottesman, I. I. (1984). *Schizophrenia and Genetic Risks.* Arlington, Va.: National Alliance for the Mentally Ill.

Gottesman, I. I. and Erlenmeyer-Kimling, L. (Eds.) (1971). Differential reproduction in invidiuals with physical and mental disorders. *Social Biology, Supplement.*

Häfner, H. and Böker, W. (1982). *Crimes of Violence by Mentally Abnormal Offenders* (trans. H. Marshall). Cambridge: Cambridge University Press.

Haller, M. H. (1963). *Eugenics — Hereditarian Attitudes in American Thought.* New Brunswick: Rutgers University Press.

*Harris, A. E. (1988) Physical disease and schizophrenia. *Schizophrenia Bulletin, 14,* 85–96.

Johnstone, E. C., Crow, T. J., Johnson, A. L. and MacMillan, J. F. (1986). The Northwick Park study of first episodes of schizophrenia: I. Presentation of the illness and problems relating to admission. *British Journal of Psychiatry, 148,* 115–120.

Kallmann, F. J. (1938). *The Genetics of Schizophrenia.* New York: Augustin.

Kevles, D. J. (1985). *In the Name of Eugenics — Genetics and the Uses of Human Heredity.* New York: Knopf.

*Krakowski, M., Volavka, J. and Brizer, D. (1986). Psychopathology and violence: A review of the literature. *Comprehensive Psychiatry, 27,* 131–148.

Larsson, T. and Sjögren, T. A. (1954). A methodological psychiatric and statistical study of a large Swedish rural population. *Acta Psychiatrica Scandinavica, Supplement 89.*

Lederberg, S. (1976). State channeling of gene flow by regulation of marriage and procreation. In A. Milunsky and G. J. Annas (Eds.), *Genetics and the Law* (pp. 247–266). New York: Plenum.

Lewis, M. S. (1989). Age incidence and schizophrenia: Part I. The season of birth controversy. *Schizophrenia Bulletin, 15,* 59–73.

Lifton, R. J. (1986). *The Nazi Doctors — Medical Killing and the Psychology of Genocide.* New York: Basic Books.

McEvoy, J. P., Hatcher, A. and Appelbaum, P. S. (1983). Chronic schizophrenic women's attitudes toward sex, pregnancy, birth control, and childbearing. *Hospital and Community Psychiatry, 34,* 536–539.

McGuffin, P. and Sturt, E. (1986). Genetic markers in schizophrenia. *Human Heredity, 36,* 65–88.

McKusick, V. A. (1990). *Mendelian Inheritance in Man: Catalogs of Autosomal Dominant, Autosomal Recessive, and X-Linked Phenotypes* (9th edition). Baltimore: Johns Hopkins University Press.

Meehl, P. E. (1990). Toward an integrated theory of schizotaxia, schizotypy, and schizophrenia. *Journal of Personality Disorders, 4:* 1–99.

Monahan, J. (1988). Risk assessment of violence among the mentally disordered: Generating useful knowledge. *International Journal of Law and Psychiatry, 11,* 249–257.

Monhahan, J. and Steadman, H. J. (1983). Crime and mental disorder: An epidemiological approach. In M. Toney and N. Morris (Eds.), *Crime and Justice: An Annual Review of Research,* Vol. 4 (pp. 145–189). Chicago: University of Chicago Press.

Müller-Hill, B. (1988). *Murderous Science* (trans. G. R. Frazer). New York: Oxford University Press.

Ødegaard, Ø. (1975). Social and ecological factors in the etiology, outcome, treatment and prevention of mental disorders. In K. P. Kisker, J. E. Meyer, C. Müller and E. Strömgren (Eds.), *Psychiatrie der Gegenwart,* Vol. 3 (pp. 151–198). Berlin: Springer-Verlag.

Paul, J. (1973). State eugenic sterilization history: A brief overview. In J. Robitscher (Ed.), *Eugenic Sterilization* (pp. 25–49). Springfield: C. C. Thomas.

*Plomin, R. (1990). The role of inheritance in behavior. *Science, 248:* 183–188.

Proctor, R. N. (1988). *Racial Hygiene — Medicine Under the Nazis.* Cambridge: Harvard University Press.

Rabkin, J. G. (1979). Criminal behavior of discharged mental patients: A critical appraisal of the research. *Psychological Bulletin, 86,* 1–27.

Reilly, P. R. (1985) Eugenic sterilization in the United States. In A. Milunsky and G. J. Annas (Eds.), *Genetics and the Law III* (pp. 227–241). New York: Plenum.

Robitscher, J. (ed.) (1973). *Eugenic Sterilization.* Springfield: C. C. Thomas.

Sartorious, N., Jablensky, A., Korten, A., Ernberg, G., Anker, M., Cooper, J. E. and Day, R. (1986). Early manifestations and first-contact incidence of schizophrenia in different cultures. *Psychological Medicine, 16,* 909–928.

Saugstad, L. F. (1989). Social class, marriage, and fertility in schizophrenia. *Schizophrenia Bulletin, 15,* 9–43.

Saugstad, L. F. and Ødegaard, Ø. (1979). Mortality in psychiatric hospitals in Norway 1950–74. *Acta Psychiatrica Scandinavica, 59,* 431–447.

Shore, D., Filson, C. R. and Johnson, W. E. (1988). Violent crime arrests and paranoid schizophrenia: The White House case studies. *Schizophrenia Bulletin, 14,* 279–281.

Slater, E. (1936). German eugenics in practice. *Eugenics Review, 27,* 285–295. Reprinted in J. Shields and I. I. Gottesman (Eds.) (1971). *Man, Mind, and Heredity — Selected Papers of Eliot Slater on Psychiatry and Genetics* (pp. 281–292). Baltimore: Johns Hopkins University Press.

Slater, E., Hare, E. H. and Price, J. (1971). Marriage and fertility of psychotic patients compared with national data. In I. I. Gottesman and L. Erlenmeyer-Kimling (Eds.), Fertility and reproduction in physically and mentally disordered individuals. *Social Biology, Supplement.*

Smith, C. (1971). Recurrence risks for multifactorial inheritance. *American Journal of Human Genetics, 23,* 578–588.

*Smith, C. and Mendell, N. R. (1974). Recurrence risks from family history and metric traits. *Annals of Human Genetics, 37,* 275–286.

Spector, T. D. and Silman, A. J. (1987). Does the negative association between rheumatoid arthritis and schizophrenia provide clues to the aetiology of rheumatoid arthritis? *British Journal of Psychiatry, 26,* 307–310.

Stevens, B. C. (1969). *Marriage and Fertility of Women Suffering from Schizophrenia or Affective Disorders.* London: Oxford University Press.

Taylor, P. J. (1982). Schizophrenia and violence. In J. Gunn and D. P. Farrington (Eds.), *Abnormal Offenders, Delinquency, and the Criminal Justice System* (pp. 269–284). New York: John Wiley & Sons.

Taylor, P. J. (1987). Social implication of psychosis. *British Medical Bulletin, 43,* 718–740.

Taylor, P. J. (1988). Forensic psychiatry. *British Journal of Psychiatry, 153,* 271–278.

Tsuang, M. T., Simpson, J. C. and Kronfol, Z. (1982). Subtypes of drug abuse with psychosis: Demographic characteristics, clinical features, and family history. *Archives of General Psychiatry, 39*, 141–147.

Vinogradov, S., Gottesman, I. I. and Moises, H. (1991). Schizophrenia and rheumatoid arthritis (unpublished manuscript).

*Vogel, F. (1987). Human genetics and the responsibility of the medical doctor. In F. Vogel and K. Sperling (Eds.), *Human Genetics* (pp. 44–53). Berlin: Springer-Verlag.

*Vogler, G. P., Gottesman, I. I., McGue, M. K., and Rao, D. C. (1990). Mixed model segregation analysis of schizophrenia in the Lindelius Swedish pedigrees. *Behavior Genetics* (in press).

*Warner, R. (1981). The influence of economic factors on outcome in schizophrenia. *Psychiatry and Social Science, 1*, 79–106.

Chapter 11

*Asarnow, R. F., Graholm, E. and Sherman, T. (1990). Span of apprehension in schizophrenia. In S. Steinhauer, J. Gruzelier and J. Zubin (Eds.), *Handbook of Schizophrenia: Neuropsychology, Psychophysiology and Information Processing*. Amsterdam: Elsevier (in press).

*Bloom, F. E. and Lazerson, A. (1988). *Brain, Mind, and Behavior* (2nd edition). New York: W. H. Freeman.

Carlson, M., Earls, F. and Todd, R. (1988). The importance of regressive changes in the development of the nervous system: Towards a neurobiological theory of child development. *Psychiatric Developments, 6*, 1–22.

Ciompi, L. (1988). Learning from outcome studies: Toward a comprehensive biological-psychosocial understanding of schizophrenia. *Schizophrenia Research, 1*, 373–384.

*Cutting, J. (1985). *The Psychology of Schizophrenia*. Edinburgh: Churchill Livingstone.

Edwards, J. H. (1969). Familial predisposition in man. *British Medical Bulletin, 25*, 58–64.

Erlenmeyer-Kimling, L. (1987). Biological markers for the liability to schizophrenia. In H. Helmchen and F. A. Henn (Eds.), *Biological Perspectives of Schizophrenia* (pp. 33–56). New York: John Wiley & Sons.

Faraone, S. V. and Tsuang, M. T. (1985). Quantitative models of the genetic transmission of schizophrenia. *Psychological Bulletin, 98*, 41–66.

Feinberg, I. (1982). Schizophrenia: Caused by a fault in programmed synaptic elimination during adolescence? *Journal of Psychiatric Research, 17,* 319–334.

Goldstein, M. J. (1987). Psychosocial issues. *Schizophrenia Bulletin, 13,* 157–171.

Gottesman, I. I. and Shields, J. (1967). A polygenic theory of schizophrenia (Abstract). *Science, 156,* 537–538. *Proceedings of the National Academy of Sciences, 58,* 199–205. Reprinted in 1972, *International Journal of Mental Health, 1,* 107–115.

Gottesman, I. I. and Shields, J. (1972). *Schizophrenia and Genetics: A Twin Study Vantage Point.* New York: Academic Press.

Gottesman, I. I. and Shields, J. (with the assistance of D. R. Hanson) (1982). *Schizophrenia: The Epigenetic Puzzle.* New York: Cambridge University Press.

*Grove, W. M. (1982). Psychometric detection of schizotypy. *Psychological Bulletin, 92,* 27–38.

Harvey, P. D. and Walker, E. F. (Eds.) (1987). *Positive and Negative Symptoms of Psychosis.* Hillsdale, N. J.: Erlbaum Assoc.

Heston, L. L. (1970). The genetics of schizophrenia and schizoid disease. *Science, 167,* 249–256.

*Hoffman, R. E. and Dobscha, S. K. (1989). Cortical pruning and the development of schizophrenia. *Schizophrenia Bulletin, 15,* 477–490.

Katschnig, H. (1987). Vulnerability and trigger models/rehabilitation: discussion. In H. Häfner, W. F. Gattaz and W. Janzarik (Eds.), *Search for the Causes of Schizophrenia* (pp. 353–358). Berlin: Springer-Verlag.

*Kerr, A. and Smith, P. (Eds.) (1986). *Contemporary Issues in Schizophrenia.* London: Gaskell.

Lalouel, J. M., Rao, D. C., Morton, N. E. and Elston, R. D. (1983). A unified model of complex segregation analysis. *American Journal of Human Genetics, 35,* 816–826.

Leff, J. and Vaughn, C. (1985). *Expressed Emotion in Families: Its Significance for Mental Illness.* New York: Guilford Press.

Liberman, R. P., Nuechterlein, K. H. and Wallace, C. J. (1982). Social skills training and the nature of schizophrenia. In J. P. Curran and P. M. Monti (Eds.), *A Practical Handbook for Assessment and Treatment* (pp. 5–56). New York: Guilford Press.

Lewontin, R. C., Rose, S. and Kamin, L. J. (1984). *Not in Our Genes.* New York: Pantheon Books.

McGue, M., Gottesman, I. I. and Rao, D. C. (1985). Resolving genetic models for the transmission of schizophrenia. *Genetic Epidemiology, 2*, 99–110.

McGue, M. and Gottesman, I. I. (1989). Genetic linkage in schizophrenia: Perspectives from genetic epidemiology. *Schizophrenia Bulletin, 15*, 281–292.

McKusick, V. A. (1990). *Mendelian Inheritance in Man: Catalogs of Autosomal Dominant, Autosomal Recessive, and X-Linked Phenotypes* (9th edition). Baltimore: Johns Hopkins University Press.

Meehl, P. E. (1962). Schizotaxia, schizotypy, schizophrenia. *American Psychologist, 17*, 827–838.

Meehl, P. E. (1972). Specific genetic etiology, psychodynamics, and therapeutic nihilism. *International Journal of Mental Health, 1*, 10–27.

Meehl, P. E. (1986). Diagnostic taxa as open concepts: Metatheoretical and statistical questions about reliability and construct validity in the grand strategy of nosological revision. In T. Millon and G. Klerman (Eds.), *Contemporary Directions in Psychopathology: Toward the DSM-IV* (pp. 215–231). New York: Guilford Press.

*Meehl, P. E. (1989). Schizotaxia revisited. *Archives of General Psychiatry, 46*, 935–944.

*Neale, J. M. and Oltmanns, T. F. (1980). *Schizophrenia*. New York: John Wiley & Sons.

Nuechterlein, K. H. (1987). Vulnerability models for schizophrenia: State of the art. In H. Häfner, W. F. Gattaz and W. Janzarik (Eds.), *Search for the Causes of Schizophrenia* (pp. 297–316). Berlin: Springer-Verlag.

Nuechterlein, K. H. and Dawson, M. E. (1984). Information processing and attentional functioning in the developmental course of schizophrenic disorders. *Schizophrenia Bulletin, 10*, 160–203.

*Robertson, F. W. (1981). The genetic component in coronary heart disease: A review. *Genetical Research, 37*, 1–16.

Saugstad, L. F. (1989). Social class, marriage, and fertility in schizophrenia. *Schizophrenia Bulletin, 15*, 9–43.

*Shaw, M. W. (1984). Presidential address: To be or not to be? That is the question. *American Journal of Human Genetics, 36*, 1–9.

Shields, J., Heston, L. L. and Gottesman, I. I. (1975). Schizophrenia and the schizoid: The problem for genetic analysis. In R. R. Fieve, D. Rosenthal and H. Brill (Eds.), *Genetic Research in Psychiatry* (pp. 167–197). Baltimore: Johns Hopkins University Press.

Vogel, F. and Motulsky, A. G. (1986). *Human Genetics*. Berlin: Springer-Verlag.

Zubin, J. (1987). Closing comments. In H. Häfner, W. F. Gattaz and W. Janzarik (Eds.), *Search for the Causes of Schizophrenia* (pp. 359–365). Berlin: Springer-Verlag.

Chapter 12

*Andreasen, N. C. (1984). *The Broken Brain: The Biological Revolution in Psychiatry*. New York: Harper & Row.

Bassett, A. S. (1989). Chromosome 5 and schizophrenia: Implications for genetic linkage studies, current and future. *Schizophrenia Bulletin, 15*, 393–402.

Baur, M. P. (1986). Genetic analysis workshop IV: Insulin dependent diabetes mellitus — summary. *Genetic Epidemiology, 1, (Supplement)*, 299–312.

Böök, J. (1953). A genetic and neuropsychiatric investigation of a north-Swedish population. *Acta Genetica et Statistica Medica (Basel), 4*, 1–100.

Botstein, D., White, R., Skolnick, M. and Davis, R. W. (1980). Construction of a genetic linkage map in man using restriction fragment length polymorphisms. *American Journal of Human Genetics, 32*, 312–331.

Buchsbaum, M. S. and Haier, R. J. (1987). Functional and anatomical brain imaging: Impact of schizophrenia research. *Schizophrenia Bulletin, 13*, 115–132.

Cloninger, C. R., Reich, T. and Yokoyama, S. (1983). Genetic diversity, genome organization, and investigation of the etiology of psychiatric diseases. *Psychiatric Developments, 3*, 225–246.

Congress of the United States, Office of Technology Assessment (1988). *Mapping Our Genes*. Baltimore: Johns Hopkins University Press.

*Donis-Keller, H. and 32 other authors (1987). A genetic linkage map of the human genome. *Cell, 51*, 319–337.

Early, T. S., Posner, M. I., Reiman, E. M., and Raichle, M. E. (1989). Left striato-pallidal hyperactivity in schizophrenia, Part II: Phenomenology and thought disorder. *Psychiatric Developments 7*, 109–121.

Erlenmeyer-Kimling, L. (1987). Biological markers for the liability to schizophrenia. In H. Helmchen and F. A. Henn (Eds.), *Biological Perspectives of Schizophrenia* (pp. 33–56). New York: John Wiley & Sons.

Field, L. L. (1988). Insulin-dependent diabetes mellitus: A model for the study of multifactorial disorders. *American Journal of Human Genetics, 43*, 793–798.

Field, L. L. (1989). Genes predisposing to insulin-dependent diabetes mellitus (IDDM) in multiplex families. *Genetic Epidemiology, 6*, 101–106.

Fish, B. (1987). Infant predictors of the longitudinal course of schizophrenic development. *Schizophrenia Bulletin, 13*, 395–409.

Friedhoff, A. J., Pickar, D., Axelrod, J., Creese, I., Davis, K. L., Gallagher, D. W., Greengard, P., Housman, D., Maas, J. W., Richelson, E., Roth, R. H. and Watson, S. J. (1989). Neurochemistry and neuropharmacology. *Schizophrenia Bulletin, 14*, 399–412.

*Frith, C. D. and Done, D. J. (1988). Towards a neuropsychology of schizophrenia. *British Journal of Psychiatry, 153*, 437–443.

*Gershon, E. S., Matthysse, S., Breakfield, X. O. and Ciaranello, R. D. (Eds.) (1981). *Genetic Research Strategies in Psychobiology and Psychiatry*. Pacific Grove: Boxwood Press.

Gilliam, T. C., Freimer, N., Powichik, P. P., Kaufman, C. A., Bassett, A. S. and Wasmuth, J. J. Characterization of a chromosome 5 deletion hybrid cell line and mapping of markers to a region which co-segregates with schizophrenia. *Genetics*, submitted 1989.

*Goldberg, T. E., Ragland, J. D., Gold, J. M., Bigelow, L. B., Torrey, E. F., and Weinberger, D. R. (1990). Neuropsychological assessment of monozygotic twins discordant for schizophrenia. *Archives of General Psychiatry* (in press).

Gottesman, I. I. and Bertelsen, A. (1989). Confirming unexpressed genotypes for schizophrenia: Risks in the offspring of Fischer's Danish identical and fraternal discordant twins. *Archives of General Psychiatry, 46*, 867–872.

Gurling, H. (1986). Candidate genes and favoured loci: Strategies for molecular genetic research into schizophrenia, manic depression, autism, alcoholism and Alzheimer's disease. *Psychiatric Developments, 4*, 289–309.

Gusella, J. F., Wexler, N. S., Conneally, P. M., Naylor, S. L., Anderson, M. A., Ranzi, R. E., Watkins, P. C., Ottina, K., Wallace, M. R., Sakaguchi, A. Y., Young, A. B., Shoulson, I., Bonilla, E. and Martin, J. B. (1983). A polymorphic DNA marker genetically linked to Hungtington's disease. *Nature, 306*, 234–238.

Hanson, D. R., Gottesman, I. I. and Heston, L. L. (1990). Long range schizophrenia forcasting: Many a slip twixt cup and lip. In J. Rolf, K. Nuechterlein, A. Masten and D. Cicchetti (Eds.), *Risk and Protective Factors in the Development of Schizophrenia*. (pp. 424–444) New York: Cambridge University Press.

Haracz, J. L. (1985). Neural plasticity in schizophrenia. *Schizophrenia Bulletin, 11*, 191–229.

*Hari, R. and Lounasmaa, O. V. (1989). Recording and interpretation of cerebral magnetic fields. *Science, 244*, 432–436.

*Holzman, P. S. (1987). Recent studies of psychophysiology in schizophrenia. *Schizophrenia Bulletin, 13*, 49–75.

Iacono, W. G., Bassett, A. S. and Jones, B. D. (1988). Eye tracking dysfunction is associated with partial trisomy of chromosome 5 and schizophrenia. *Archives of General Psychiatry, 45,* 1140–1141.

*Innis, M. A., Gelfand, D. H., Sninsky, J. J. and White, T. J. (Eds) (1989). *PCR Protocol: A Guide to Methods and Applications.* New York: Academic Press.

Keith, S. J. and Matthews, S. M. (1988). *A National Plan for Schizophrenia Research: Panel Recommendations.* Rockville: NIMH (Reprinted as *Schizophrenia Bulletin, 14(3),* 1988).

Kennedy, J. L., Giuffra, L. A., Moises, H. W., Cavalli-Sforza, L. L., Pakstis, A. J., Kidd, J. R., Castiglione, C. M., Sjogren, B., Wetterberg, L. and Kidd, K. K. (1988). Evidence against linkage of schizophrenia to markers on chromosome 5 in a northern Swedish pedigree. *Nature, 336,* 167–170.

Kennedy, J. L., Wetterberg, L., Sjogren, B., Giuffra, L. A., Pakstis, A. J. and Kidd, K. K. (1989). Genetic heterogeneity in schizophrenia: Contributions from a north Swedish kindred (Abstract). *Schizophrenia Research, 2,* 43.

*Lander, E. S. and Botstein, D. (1989). Mapping Mendelian factors underlying quantitative traits using RFLP linkage maps. *Genetics, 121,* 185–199.

*Lipp, H. P. (1989). Non-mental aspects of encephalisation: the forebrain as a playground of mammalian evolution. *Human Evolution, 4,* 45–53.

McGuffin, P. and Sturt, E. (1986). Genetic markers in schizophrenia. *Human Heredity, 36,* 65–88.

McGuffin, P., Sargeant, M., Hett, G., Tidmarsh, S., Whatley, S., and Marchbanks, R. M. (1990). Exclusion of a schizophrenia susceptibility gene from the chromosome 5q11-q13 region. New data and a reanalysis of previous reports. *American Journal of Human Genetics* (in press).

McKenna, P. J. (1987). Pathology, phenomenology and the dopamine hypothesis of schizophrenia. *British Journal of Psychiatry, 151,* 288–301.

McKusick, V. A. (1990). *Mendelian Inheritance in Man: Catalogs of Autosomal Dominant, Autosomal Recessive, and X-Linked Phenotypes* (9th edition). Baltimore: Johns Hopkins University Press.

Mednick, S. A., Cannon, T., Parnas, J., and Schulsinger, F. (1989). 27 year follow-up of the Copenhagen high-risk for schizophrenia project: Why did some of the high-risk offspring become schizophrenic? (Abstract). *Schizophrenia Research, 2,* 14.

Meltzer, H. Y. (1987). Biological studies in schizophrenia. *Schizophrenia Bulletin, 13,* 77–111.

Moldin, S., Gottesman, I. I., Erlenmeyer-Kimling, L., and Cornblatt, B. A. (1990). Psychometric deviance in offspring at risk for schizophrenia. *Psychiatry Research* (in press).

National Advisory Mental Health Council Report to Congress on the Decade of the Brain (1988). *Approaching the 21st Century: Opportunities for NIMH Neuroscience Research.* Rockville: NIMH (DHHS Publication No. [ADM] 89-1580.)

*Olson, L., Stromberg, I., Bygdeman, M., Granholm, A.-C., Hoffer, B., Freedman, R. and Seiger, Å. (1987). Human fetal tissues grafted to rodent hosts: Structural and functional observations of brain, adrenal and heart tissues in oculo. *Experimental Brain Research, 67*, 163–178.

*Pardo, J. V., Pardo, P. J., Janer, K. W., and Raichle, M. E. (1990). The anterior cingulate cortex mediates processing selection in the Stroop attentional conflict paradigm. *Proceedings of the National Academy of Sciences USA, 87*, 256–259.

*Patterson, A. H., Lander, E. S., Hewitt, J. D., Peterson, S., Lincoln, S. E. and Tanksley, S. D. (1988). Resolution of quantitative traits into Mendelian factors by using a complete linkage map of restriction fragment length polymorphisms. *Nature, 335*, 721–726.

Pearson, J. S. and Kley, I. B. (1957). On the application of genetic expectancies as age specific base rates in the study of human behavior disorders. *Psychological Bulletin, 54*, 406–420.

*Posner, M. I., Early, T. S., Reiman, E., Pardo, P. and Dhawan, M. (1988). Asymmetries in hemispheric control of atention in schizophrenia. *Archives of General Psychiatry, 45*, 814–821.

*Posner, M. I. and Early, T. S. (1991-in press). Schizophrenia and the development of attention. *Transmission* (in press).

*Ravich-Shcherbo, I. V. (1988). [*The Role of Environment and Heredity in the Foundation of the Individuality of the Person.*] Moskva: Pedagogika.

Resnick, S. M., Gur, R. E., Torrey, E. F., Mozley, P. D., Taleff, M. M., Muehllehner, G., Gur, R. C., Gottesman, I. I., Reivich, M. and Alavi, A. (1989). PET scan studies: Initial results in identical twins discordant for schizophrenia. *Schizophrenia Research, 2*, 255.

Risch, N. (1990). Genetic linkage and complex diseases, with special reference to psychiatric disorders. *Genetic Epidemiology, 7*, 3–16.

Rolf, J., Nuechterlein, K., Masten, A. and Cicchetti, D. (Eds.) (1990). *Risk and Protective Factors in the Development of Schizophrenia.* New York: Cambridge University Press.

St. Clair, D., Blackwood, D., Muir, W., Bailie, D., Hubbard, A., Wright, A. and Evans, H. J. (1989). No linkage of chromosome 5q11-q13 markers to schizophrenia in Scottish families. *Nature, 339*, 305–309.

Seeman, P., Farde, L. and Sedvall, G. (1988). Brain dopamine receptors in schizophrenia: PET problems. *Archives of General Psychiatry, 45*, 598–600.

Sherrington, R., Brynjolfsson, J., Petursson, H., Potter, M., Dudleston, K., Barraclough, B., Wasmuth, J., Dobbs, M. and Gurling, H. (1988). Localization of a susceptibility locus for schizophrenia on chromosome 5. *Nature, 336*, 164–170.

*Sing, C. F., Boerwinkle, E., Moll, P. P. and Templeton, A. R. (1988). Characterization of genes affecting quantitative traits in humans. In B. S. Weir, E. J. Eisen, M. M. Goodman and G. Namkoong (Eds.), *Proceedings of the Second International Conference on Quantitative Genetics* (pp. 250–269). Sunderland, Mass.: Sinauer Associates.

*Snyder, S. H. and Largent, B. L. (1989). Receptor mechanisms in antipsychotic drug action: Focus on sigma receptors. *Journal of Neuropsychiatry and Clinical Neurosciences, 1*, 7–15.

*Sturt, E. and McGuffin, P. (1985). Can linkage and marker association resolve the genetic aetiology of psychiatric disorders? Review and argument. *Psychological Medicine, 15*, 455–462.

Suddath, R. L., Christison, G. W., Torrey, E. F., Casanova, M. F., and Weinberger, D. R. (1990). Anatomical abnormalities in the brains of monozygotic twins discordant for schizophrenia. New England Journal of Medicine, 322:789–794.

Tune, L. E., Wong, D. F., Pearlson, G. D., Young, L. T., Villemagne, V., Fannals, R. F., Young, D., Wilson, A. A., Ravert, H. T., Links, J. M., Midha, K., Wagner, H. N. and Jedde, A. (1989). D2 dopamine receptors in drug naive schizophrenics: Update on 20 subjects (Abstract). *Schizophrenia Research, 2*, 114.

*Vartanian, M. E. (Ed.) (1988). *Neuronal Receptors, Endogenous Ligands and Biotechnical Approaches.* Madison, CT: International Universities Press.

*Waddington, J. L. (1989). Sight and insight: Brain dopamine receptor occupancy by neuroleptics visualized in living schizophrenic patients by positron emission tomography. *British Journal of Psychiatry, 154*, 433–436.

Wagner, H. N., Weinberger, D. R., Kleinman, J. E., Casanova, M. F., Gibbs, C. J., Gur, R. E., Hornykiewicz, O., Huhar, M. J., Pettegrew, J. W. and Seeman, P. (1988). Neuroimaging and neuropathology. *Schizophrenia Bulletin, 14*, 383–397.

Watt, N. F., Anthony, E. J., Wynne, L. C. and Rolf, J. E. (1984). *Children at Risk for Schizophrenia: A Longitudinial Perspective.* New York: Cambridge University Press.

*Weeks, D. E. and Lange, K. (1988). The affected-pedigree-member method of linkage analysis. *American Journal of Human Genetics, 42*, 315–326.

Weinberger, D. and Kleinman, J. (1986). Observations on the brain in schizophrenia. *Annual Review Psychiatry Update, 5*, 42–67.

*Weir, B. S., Eisen, E. J., Goodman, M. M. and Namkoong, G. (1988). *Proceedings of the Second International Conference on Quantitative Genetics*. Sunderland, Mass.: Sinauer Associates.

Wong, D. F., Pearlson, G. D., Tune, L. E., Young, C., Ross, C., Villemagne, V., Dannals, R. F., Young, D., Parker, R., Wilson, A. A., Ravert, H. T., Links, J., Midha, K., Wagner, H. N. and Gjedde, A. (1989). Update on PET methods for D2 dopamine receptors in schizophrenia and bipolar disorder (Abstract). *Schizophrenia Research, 2*, 115.

Name Index

Subject Index

About the Author

Irving I. Gottesman is one of the world's leading authorities on the origins of schizophrenia. He is the winner of numerous honors and awards including the 1990 Dobzhansky Award for Lifelong Contributions to Behavioral Genetics. Gottesman received his Ph.D. from the University of Minnesota and is presently Commonwealth Professor of Psychology and Professor of Pediatrics at the University of Virginia. He was recently elected Honorary Fellow of the Royal College of Psychiatrists.